SpringerBriefs in Latin American Studies

SpringerBriefs in Latin American Studies promotes quality scientific research focusing on Latin American countries. The series accepts disciplinary and interdisciplinary titles related to geographical, environmental, cultural, economic, political and urban research dedicated to Latin America. The Series will publish compact volumes (50 to 125 pages) refereed by a region or country expert specialized in Latin American studies. We offer fast publication time of 8 to 12 weeks after acceptance. The series aims to raise the profile of Latin American studies, showcasing important works developed focusing on the region. It is aimed at researchers, students, and everyone interested in Latin American topics.

More information about this series at http://www.springer.com/series/14332

Eduardo Salazar Moreira ·
Marcela Palomino-Schalscha

Road Expansion in the Peruvian Amazon

The 'Enchantments' of the Manu Road

CLAG
clagscholar.org

Springer

Eduardo Salazar Moreira
School of Geography
Environment and Earth Sciences
Victoria University of Wellington
Wellington, New Zealand

Marcela Palomino-Schalscha
School of Geography
Environment and Earth Sciences
Victoria University of Wellington
Wellington, New Zealand

ISSN 2366-763X ISSN 2366-7648 (electronic)
SpringerBriefs in Latin American Studies
ISBN 978-3-030-47181-1 ISBN 978-3-030-47182-8 (eBook)
https://doi.org/10.1007/978-3-030-47182-8

This Springer imprint is published by the registered company Springer Nature Switzerland AG
The registered company address is: Gewerbestrasse 11, 6330 Cham, Switzerland

For my Aranza, my Luz.

—Marcela

Eduardo's Acknowledgements

I would like to thank everyone that facilitated my fieldwork in Manu, especially Chaskawasi Manu for allowing me to stay at their facilities, the Crees Foundation for their support with boat transportation, the communities that allowed me to conduct research with them and the people of the Manu province, whom I remember lovingly. I am also deeply grateful for the help that Dr. Marisa Wilson, Dr. Kanchana Ruwanpura and Dr. Samantha Staddon, from the University of Edinburgh, provided during my M.Sc. studies, which were the first step in what will hopefully become a meaningful career as an academic. Of course, I would not have survived this process without the continued support and patience of my family and loved ones, who have provided unconditional encouragement and patient understanding. Last but definitely not least, I would like to thank my co-author and awesome Ph.D. supervisor Marcela, who bravely embarked on this adventure as soon as I proposed it.

Contents

About the Authors

Eduardo Salazar Moreira is a Ph.D. candidate at the School of Geography, Environment and Earth Sciences at Victoria University of Wellington in New Zealand. After two years working for a not-for-profit organization in the Manu province of Peru's southern Amazon, Eduardo began his research about the Manu Road as part of the M.Sc. in Environment and Development at the University of Edinburgh in Scotland. His Master's dissertation is the basis for this book and the research he will continue to conduct through his Ph.D. studies.

Marcela Palomino-Schalscha is a Lecturer in Geography and Development Studies at Victoria University of Wellington in New Zealand. Her research interests lie at the intersection of social geography, development studies and political ecology, with a special emphasis on Indigenous rights. Most of her work is located in Latin America, where she theorizes the politics of scale and place, diverse and solidarity economies, decolonisation, identity politics, Indigenous tourism and relational ontologies. More recently, she has also embarked on the use of *arpilleras*, textiles with political content, as more-than-textual research methods to explore the experience of refugee-background and migrant Latin American women in New Zealand. She is the co-editor of *The Routledge Handbook of Latin American Development* (Cupples, J., Palomino-Schalscha, M., & Prieto, M. (Eds.), 2018, Routledge), and *Indigenous Places and Colonial Spaces: The Politics of Intertwined Relations* (Gombay, N., & Palomino-Schalscha, M., 2018, Routledge). She is also Co-editor of ACME: An International E-Journal for Critical Geographies.

Abbreviations

ACR	Amarakaeri Communal Reserve (*Reserva Comunal Amarakaeri*)
AIDESEP	Inter-ethnic Development Association of Peruvian Amazonia (*Asociación Interétnica de Desarrollo de la Amazonía Peruana*)
AMD River	Alto Madre de Dios River (*Río Alto Madre de Dios*)
COHARYIMA	Harakmbut, Yine and Matsiguenka Council (*Consejo Harakmbut, Yine y Machiguenga*)
FENAMAD	Native Federation of the Madre de Dios River and Effluents (*Federación Nativa del Río Madre de Dios y Afluentes*)
GOREMAD	Madre de Dios Regional Government (*Gobierno Regional de Madre de Dios*)
IIRSA	Initiative for the Integration of South American Regional Infrastructure (*Iniciativa para la Integración de la Infraestructura Regional Suramericana*)
MAA	Manu Agricultural Agency (*Agencia Agraria Manu*)
MEF	Ministry of Economics and Finance (*Ministerio de Economía y Finanzas*)
MINAM	Ministry of the Environment (*Ministerio del Ambiente*)
MNP	Manu National Park (*Parque Nacional del Manu*)
MTC	Ministry of Transport (*Ministerio de Transportes y Comunicaciones*)
SERNANP	National Parks Service (*Servicio Nacional de Áreas Protegidas por el Estado*)
SRG	Manu Sub-Regional Government (*Sub-Región Manu*)

Abbreviations that are commonly used in Spanish have been used in this book in their original versions. Abbreviations that are not commonly used in Spanish are based on the translation of their names in English.

List of Figures

Chapter 1
Introduction

Abstract The aspirations of development that forthcoming roads generate in the Amazonian frontier take a strong hold on geographically, politically and economically isolated communities, despite the potential for failure of these infrastructures. These aspirations clash with conservation interests in the Manu province of the Peruvian Amazon, where the first road to ever reach this province is being gradually expanded in between the two protected areas that occupy most of its territory. In this introductory chapter we present the issues surrounding this road expansion project by providing an initial overview of the situation in both local and broader terms. We then briefly discuss the relevance of this case study for broader debates around development, conservation and Indigenous rights, and introduce our approach to unpack these complexities. Our review of the situation in Manu through ethnographic accounts later allows us to conclude that the territorial pressures driven by the establishment of protected areas and by the expansion of regional road networks significantly influence popular support for the expansion of the Manu Road.

Keywords Manu road · First roads · Frontier infrastructure · Enchantments · Indigenous · Amazonia · Political ecology · Conservation · Territoriality · Development

1.1 The Case of the Manu Road, a Biologist's Paradise from a Political Ecology Perspective

The aspirations of development that forthcoming roads generate in the Amazonian frontier take a strong hold on communities that feel isolated from the rest of the countries they are in and their respective economies (Harvey and Knox 2012, 2015). Despite the anticipation for the positive effects of roads, it is the very nature of the frontier, the 'friction' that meets universal notions of modernity in the far reaches of state authority and the capitalist economy, what often prevents roads from generating the anticipated changes (Harvey and Knox 2012, 2015; Tsing 2005). As proposed by Harvey and Knox, however, the failure of roads to generate the expected development

E. Salazar Moreira and M. Palomino-Schalscha, *Road Expansion in the Peruvian Amazon*, SpringerBriefs in Latin American Studies, https://doi.org/10.1007/978-3-030-47182-8_1

outcomes exacerbates the feelings of need for these infrastructures and 'enchants' off-road communities, making their construction an imperative in the local imaginary (2012).

Nonetheless, in the Manu province of Peru's Madre de Dios region, where the Manu National Park (MNP) and the adjacent Amarakaeri Communal Reserve (ACR) occupy most of the territory, isolation has been an asset for conservation (Gallice et al. 2017; INRENA 2008; SERNANP 2013, 2014; Shepard et al. 2010). Particularly, the MNP is considered the highest pride of the Peruvian conservation sector, given its history as the first national park in the country, the presence of several Indigenous groups and its record-breaking biodiversity levels (Gallice et al. 2017; SERNANP 2013, 2014; Shepard et al. 2010). This book explores the controversies and multifaceted reasons that different actors articulate to support or resist a proposal to expand the Manu Road, the only road in this part of the Manu province (Gallice et al. 2017; Larrea-Gallegos et al. 2016). In doing so, we bridge natural and social sciences concerns that range from conservation to anthropological studies. In fact, despite the wealth of research conducted in the MNP on biological sciences and anthropology, an interdisciplinary approach to the implications of the area's protected status is largely absent (Ráez Luna 2017; Shepard et al. 2010).

The 'Manu Road',[1] the first and still the only unpaved track that penetrates into the central Manu province, has come to be known as such after many varying denominations that go from 'Regional Integration Highway' to 'economic access route', and 'local track' (Congreso de la República 2013; Gallice et al. 2017; SERNANP 2017). The debate surrounding the expansion of this infrastructure appears as a symbol of the largely apolitical approach to the management of the surrounding protected areas (Ráez Luna 2017; Robbins 2004; Shepard et al. 2010). The conflicts between conservation and modernization that road expansion near protected areas presents, raise important questions around development, the environment and power in the Amazon (Figallo and Vergara 2014; Larsen 2016). As such, it represents a characteristic Political Ecology puzzle, which calls for a deep analysis of its political causes as an attempt to envision fairer solutions to complex problems (Adams and Hutton 2007; Bryant and Bailey 2000; Escobar 2008; Robbins 2004). So far, the few studies produced regarding the Manu Road have focused on the environmental consequences of the construction and the presence of the road, and marginally on its environmentally meaningful social consequences (Gallice et al. 2017; Larrea-Gallegos et al. 2016). Identifying these consequences is certainly valuable but, considering that it is well established in the literature that road building is pernicious to rainforest ecosystems and to frontier communities, such an apolitical approach tends to prompt merely palliative solutions (Araujo et al. 2011; Arima et al. 2005, 2013; Bryant and Bailey 2000; Carvalho et al. 2002; Gallice et al. 2017; Pieck 2011; Robbins 2004; Urrunaga et al. 2012, 2018; Vergara et al. 2014).

On the other hand, taking a closer look at the forces that shape a specific road inserted in a complex infrastructural system through a Political Ecology lens can

[1]The term 'Manu Road' is used throughout this publication to refer to the road in question, and was used originally by Gallice et al. (2017).

reveal the manifold processes it involves and contribute to the identification of more nuanced, socially inclusive and just alternatives (Bryant and Bailey 2000; Harvey et al. 2017; Robbins 2004). Moreover, because roads are a key feature that shapes space for human society, the study of frontier roads like the one in Manu provides a useful insight into the types of political, social and economic relations that take place in the frontier (Uribe 2017). Studying these infrastructures can also shed light on 'processes and politics that will be of relevance to infrastructural projects more generally' (Harvey and Knox 2015, p. 15). Furthermore, adopting a Political Ecology perspective enables us to focus on the influence of power and politics upon the environment when analysing the impact of such new infrastructures (Bryant and Bailey 2000; Robbins 2004). These impacts are widespread in the region, given the impending growth of the Peruvian and South American road networks through the 'Initiative for the Integration of South American Regional Infrastructure'[2] (IIRSA) and connected projects (Barrantes et al. 2014; Barrantes and Glave 2014; Gallice et al. 2017; Kanai 2016; Zibechi 2006). The expansion of the road network prompted by IIRSA has incentivized the emergence of several 'first road' projects presently being proposed for various other areas of the Peruvian Amazon and beyond, which are currently only accessible by river (Congreso de la República 2017; SPDA 2017, 2018a, b; Zibechi 2006). Additionally, these new access routes proposed throughout the Amazonian frontier will keep playing a decisive role in the future of the conservation of protected areas like the MNP and the ACR (Gallice et al. 2017; Oliart and Biffi 2010; Ráez Luna 2017). The specific case of the Manu road is crucial given the high regard of the MNP in the Peruvian conservation landscape (Gallice et al. 2017; Shepard et al. 2010).

Thus, through an analysis of the Political Ecology of the Manu Road, current tensions between modernization, conservation and Indigenous people's rights and aspirations will be explored. In particular, the processes by which first roads are conceived and laid out in the twenty-first century Amazonian frontier will be better understood. Specifically, by delving deeper into Harvey and Knox's concept of 'the enchantments of infrastructure', this research further elucidates the way in which aspirations for modernizing roads are generated in this context (2012). Doing so also reveals the influence that power relations in the frontier have in how roads are built, used and contested (Harvey and Knox 2012). Additionally, considering that most of the current literature deals with already existing unpaved roads being paved, this research will also complement the scarce literature on 'first roads' that is available. Taking an ethnographic approach, in this book we provide an in-depth account of the messy, intertwined and politically charged ways in which 'the enchantments of infrastructure' play out in the proposal to extend the Manu Road (Harvey and Knox 2012). Our Political Ecology stance also allows us to observe how these 'enchantments' are ultimately exacerbated by the power struggle between conservation, economic and cultural interests articulated across scales and places. Therefore, the exploration of this particular case reveals the influence of elements seldom considered in cases of road expansion near protected areas (Adams and Hutton 2007; Bluwstein and Lund

[2] *Iniciativa para la Integración de la Infraestructura Regional Suramericana*, IIRSA.

2018; Delaney 2005; Raycraft 2018; Robbins 2004). As a result of our analysis, we propose that popular support for the extension of the Manu Road is significantly influenced by the results of the struggle between the forms of territorialization driven by the establishment of protected areas and by the expansion of regional road networks. These territorializing powers constrain the territoriality of local communities, increasing the power of the 'enchantments of infrastructure' near protected areas.

1.2 Book Structure and Our Use of a Narrative Style

This introductory chapter provides a first glance at the significance of the Manu Road in a territory of such natural, social and cultural complexity, as well as explaining the way in which we have analysed this case. In the next chapter, we further uncover these intricacies by exploring the historical events that form the backdrop to the context of and conflicts around the construction of the Manu Road. Chapter 3 provides a revision of relevant literature, including theoretical considerations about the social context and an analysis of 'the enchantments of infrastructure', as well as delving into key notions of territory, power and coloniality relevant to this case. Chapter 4 explains the methodological approach we have taken, as well as the methods through which this research was conducted. Based on detailed ethnographic accounts, Chaps. 5 and 6 explore the way in which 'the enchantments of infrastructure' take place in the case of the Manu Road, while Chap. 7 explains the implications of Manu's territorial issues and local power structures in the formation of these 'enchantments' (Harvey and Knox 2012). The final chapter closes this work by further clarifying the links between conservation and road construction in the area.

Chapters 5–7 include ethnographic material narrated by Eduardo. We have included these accounts to ground our discussions on the lived realities of those affected by the expansion of the Manu Road. This will allow our analysis to become more relatable to the reader (Maggio 2014), to provide a clearer perspective regarding the causes of the conflict surrounding road expansion (Aunger 1995) and to help understand the power structures that underlie it (Cameron 2012). Based on Eduardo's experience conducting fieldwork as part of his Master's dissertation research between April and June 2017 (Salazar Moreira 2017), these stories, as well as other ethnographic material included as part of our discussion, provide a glance of the complex, multifaceted and challenging realities faced by those that either promote or oppose the extension of the Manu Road.

The places Eduardo visited were accessible through the Manu Road, which is the only way to access the province by land, or through the Alto Madre de Dios (AMD) River, a waterway that connects off-road communities in the area. The original place names, names of organizations and of Indigenous groups (Yine, Matsiguenka, Harakmbut/Amarakaeri) have been used to provide a concrete context but all personal names mentioned, except those of public figures, are pseudonyms. The stories we have included have been written in first-person singular by Eduardo to emphasize their embodied, grounded nature, but our discussions are written in the first-person plural to recognize our co-authorship.

References

Adams WM, Hutton J (2007) People, parks and poverty: political ecology and biodiversity conservation. Conserv Soc 5(2):147–183

Araujo C, Bonjean CA, Combes J, Araujo C, Bonjean CA, Combes J, Motel PC (2011). Does land tenure insecurity drive deforestation in the Brazilian Amazon? CERDI, Etudes et Documents 2010(13):2–34

Arima EY, Walker RT, Souza C, Pereira R, do Canto O (2013) Spontaneous colonization and forest fragmentation in the central Amazon basin. Ann Assoc Am Geograph 103(6):1485–1501

Arima EY, Walker RT, Perz SG, Caldas M, Arima EY, Walker RT ... Caldas M (2005) Loggers and forest fragmentation: behavioral models of road building in the Amazon basin. Ann Assoc Am Geograph 95(3):525–541

Aunger R (1995) On ethnography: storytelling or science? Curr Anthropol 36(I):97–130

Barrantes R, Glave M (2014) Introducción. In: Barrantes R, Glave M (eds) Amazonía Peruana y Desarrollo Económico. GRADE, IEP, Lima, pp 13–20

Barrantes R, Fiestas J, Hopkins Á (2014) Evolución de la infraestructura de Transporte y Energía en la Amazonía Peruana. In: Barrantes R, Glave M (eds) Amazonía Peruana y Desarrollo Económico. GRADE, IEP, Lima, pp 109–160

Bluwstein J, Lund JF (2018) Territoriality by conservation in the Selous-Niassa corridor in Tanzania. World Dev 101:453–465

Bryant R, Bailey S (2000) Third world political ecology. Routledge, London and New York

Cameron E (2012) New geographies of story and storytelling. Prog Hum Geogr 36(5):573–592

Carvalho GO, Nepstad D, McGrath D, Diaz MDV, Santilli M, Barros AC (2002) Frontier expansion in the Amazon: balancing development and sustainability. Environ: Sci Policy Sustain Dev 44(3):34–44

Congreso de la República (2013) Proyecto de ley que Declara de Necesidad Pública e Interés Nacional la Construcción de la Carretera de Integración Regional de Madre de Dios

Congreso de la República (2017) Proyecto de Ley que Declara de Necesidad Pública e Interés Nacional la Construcción de la Carretera de Integración Regional Madre de Dios. Tramo Puerto Maldonado—Mazuco—Limonchayoc—Gamitana—Salvación

Delaney D (2005) Territory: a short introduction. Blackwell Publishing, Oxford

Escobar A (2008) Territories of difference: place, movements, life, redes. Duke University Press, Durham and London

Figallo M, Vergara K (2014) La Amazonía Peruana de Hoy. In: Barrantes R, Glave M (eds) Amazonía Peruana y Desarrollo Económico. GRADE, IEP, Lima, pp 47–108

Gallice G, Larrea-Gallegos G, Vásquez-Rowe I (2017) The threat of road expansion in the Peruvian Amazon. Oryx 1–9

Harvey P, Knox H (2012) The enchantments of infrastructure. Mobilities 7(4):521–536

Harvey P, Knox H (2015) Roads: an anthropology of infrastructure and expertise. Expertise, Ithaca, NY

Harvey P, Jensen CB, Morita A (2017) Introduction: infrastructural Complications. Infrastructures and social complexity: a companion. Routledge, London, pp 1–42

INRENA (2008) Plan Maestro de la Reserva Comunal Amarakaeri 2008–2012. Ministerio de Agricultura, Lima

Kanai JM (2016) The pervasiveness of neoliberal territorial design: cross-border infrastructure planning in South America since the introduction of IIRSA. Geoforum 69:160–170

Larrea-Gallegos G, Vazquez-Rowe I, Gallice G (2016) Life cycle assessment of the construction of an unpaved road in an undisturbed tropical rainforest area in the vicinity of Manu National Park, Peru. Int J Life Cycle Assess 11(17/2016):1–16

Larsen PB (2016) Derechos indígenas, gobernanza ambiental y recursos en la Amazonía peruana, Hacia una antropología de la posfrontera. Instituto de Estudios Peruanos, Lima

Maggio R (2014) The anthropology of storytelling and the storytelling of anthropology. J Compar Res Anthropol Sociol 5(2):89–106

Oliart P, Biffi V (2010) Territorialidad indígena, conservación y desarrollo: Discursos sobre la biodiversidad en la Amazonía peruana. Instituto del Bien Común, Lima
Pieck SK (2011) Beyond postdevelopment: civic responses to regional integration in the Amazon. J Cult Geogr 28(1):179–202
Ráez Luna E (2017) Pueblos Indígenas y Conservación de la Naturaleza: ¿Debates teóricos o urgencias vitales? El caso del Parque Nacional del Manu. In: Ráez Luna E (ed) Derechos Ancestrales y Conservación de la Naturaleza en Debate: El caso de los Pueblos Indígenas del Parque Nacional del Manu. Seminario Permanente de Investigación Agraria (SEPIA), Lima
Raycraft J (2018) Circumscribing communities: marine conservation and territorialization in southeastern Tanzania. Geoforum 100(2019):128–143
Robbins P (2004) Political ecology: a critical introduction. Blackwell Publishing, Oxford
Salazar Moreira E (2017) The enchantments of roads near protected areas in the Peruvian Amazon. MSc Dissertation, University of Edinburgh
SERNANP (2013) Parque Nacional del Manu, 40 años (SERNANP). Biblios, Lima
SERNANP (2014) Plan Maestro del Parque Nacional del Manu 2013–2018
SERNANP (2017) Oficio N° 620–2017—SERNANP—DGANP. Ministerio del Ambiente, Lima
Shepard GH, Rummenhoeller K, Ohl-Schacherer J, Yu DW (2010) Trouble in Paradise: indigenous populations, anthropological policies, and biodiversity conservation in Manu National Park, Peru. J Sustain Forest 29(2–4):252–301
SPDA (2017) Congresistas impulsan carreteras que perjudicarían a indígenas en aislamiento. Sociedad Peruana de Derecho Ambiental Website http://www.actualidadambiental.pe/?p=46670
SPDA (2018a) Indígenas en contra de ley que promueve carreteras en la Amazonía. Sociedad Peruana de Derecho Ambiental Website http://www.actualidadambiental.pe/?p=48424
SPDA (2018b) Nuevo gobernador de Ucayali propone carreteras perjudiciales para áreas protegidas e indígenas en aislamiento. Sociedad Peruana de Derecho Ambiental Website http://www.actualidadambiental.pe/?p=51805
Tsing AL (2005) Friction: an ethnography of global connection. Princeton University Press, New Jersey
Uribe S (2017) Frontier road: power, history and the everyday state in the Colombian Amazon. Wiley, New Jersey
Urrunaga J, Johnson A, Orbegoso I, Mulligan F (2012) La máquina lavadora: Cómo el Fraude y la Corrupción en el sistema de concesiones están Destruyendo el Futuro de los bosques del Perú. Environmental Investigation Agency, London and Washington, DC
Urrunaga J, Johnson A, Orbegozo D (2018) El Momento de la Verdad: Oportunidad o Amenaza para la Amazonía Peruana en la lucha contra el Comercio de la Madera Ilegal. Environmental Investigation Agency, London and Washington, DC
Vergara K, Figallo M, Glave M (2014) Infraestructura en la Amazonía Peruana: una propuesta para proyectar cambios en la cobertura boscosa en la carretera Pucallpa-Cruzeiro do Sul. In: Barrantes R, Glave M (eds) Amazonía Peruana y Desarrollo Económico. GRADE, IEP, Lima, pp 161–208
Zibechi R (2006) IIRSA: la integración a la medida de los mercados. Ecología Política 31:19–25

Chapter 2
The Manu Area of the Peruvian Amazon, Past and Present

Abstract In this chapter we revise the history of the Manu area, beginning from its past as a remote and inaccessible area and continuing through its recent history as a site of resource extraction and increased migration, to finally become an area of high regard for Peruvian conservation. Then, we describe the process of gradual road expansion that has continued to increase access into this yet remote place of the Peruvian Amazon. We conclude this historical review by providing details of the recent controversy surrounding the expansion of the Manu Road, which form the foundations to the analysis developed in subsequent chapters. Refer to Map 2.1 in this chapter to locate most of the places along the Manu Road and the Alto Madre de Dios river that are mentioned in the following historical review, and in our discussions and stories in the following chapters.

Keywords Manu road · Manu National Park · Amarakaeri Communal Reserve · Internal colonization · Frontier · Conservation

2.1 The History of Manu, Territorial Changes in the Inka Jungle

In this section, we recount the history of Manu province[1] that precedes the current stage of road expansion and the ensuing controversy. We detail the gradually increasing presence of non-Indigenous communities in the area and the incentives that attracted them, continuing with the establishment of protected areas as a reaction to the environmental impact of their presence, as well as the effects of these changes on the Indigenous population. This is later complemented by a more detailed account of the recent conflicts generated by the latest expansion of the Manu Road.

It is well documented that neither Incas nor Spaniards were able to penetrate the Madre de Dios river basin given the resistance of the natives to the presence of outsiders and its challenging geography of impenetrable rainforests and treacherous rivers (Gallice et al. 2017; Greene 2009; INRENA 2008; Shepard et al. 2010). Even

[1]Refer to Fig. 2.1 for a timeline of the main events described here.

E. Salazar Moreira and M. Palomino-Schalscha, *Road Expansion in the Peruvian Amazon*, SpringerBriefs in Latin American Studies, https://doi.org/10.1007/978-3-030-47182-8_2

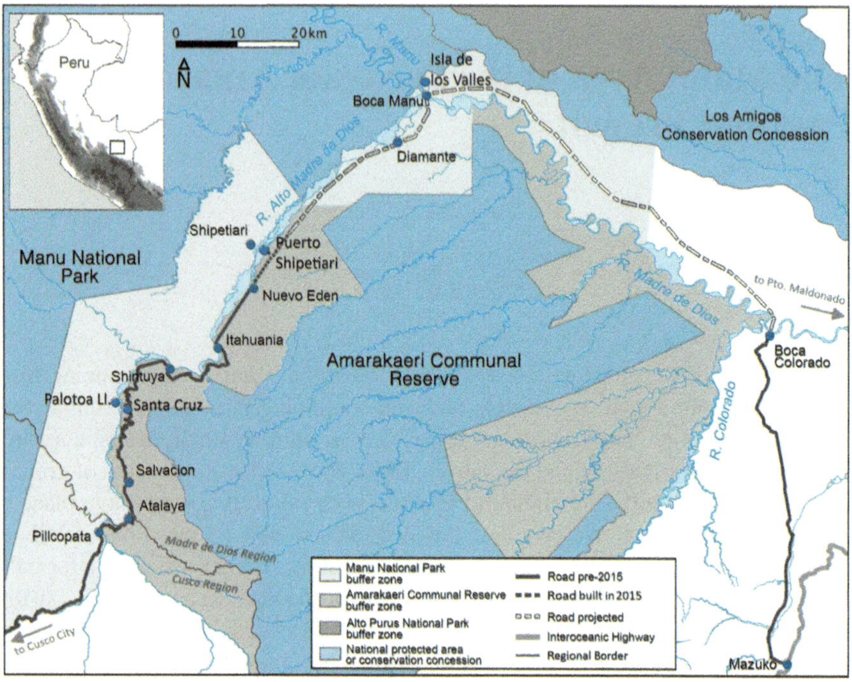

Map 2.1 The central area of Manu province: The Manu Road, the Manu National Park, the Amarakaeri Communal Reserve and the surrounding area

in the early republican times, the whole of the Peruvian Amazon was mostly disconnected from the rest of the country, from the elites in Lima and other coastal urban areas, and from the Andean highlands that stood in between (Ames 2010; Morel 2014). Despite that Cusco, the capital of the Inka Empire, was relatively nearby, trade routes halted at the edge of the Kosñipata valley, at the headwaters of the Madre de Dios river, the very end of Incan territory, currently the border between the Cusco and the Madre de Dios regions (see Map 2.1) (Gallice et al. 2017).

The rainforests of Manu have always been a symbol of remoteness, inaccessibility and mystery in the Peruvian Amazon (Gallice et al. 2017; Shepard et al. 2010). It was not until the end of the nineteenth century that foreigners, motivated by the exploitation of rubber, were able to penetrate the area (Shepard et al. 2010). But even then, they avoided the AMD river by entering Manu through the northwest, crossing an isthmus on foot from a tributary of the Ucayali basin (Shepard et al. 2010). However, after the 'rubber boom' busted a few decades later, the few Westerners that remained were mostly Christian missionaries attempting to convert the 'savage' natives, the Matsiguenkas, the Yines and the Harakmbut, who lived near the Manu and AMD rivers (Ames 2010; Gallice et al. 2017; Harvey and Knox 2012; Shepard et al. 2010).

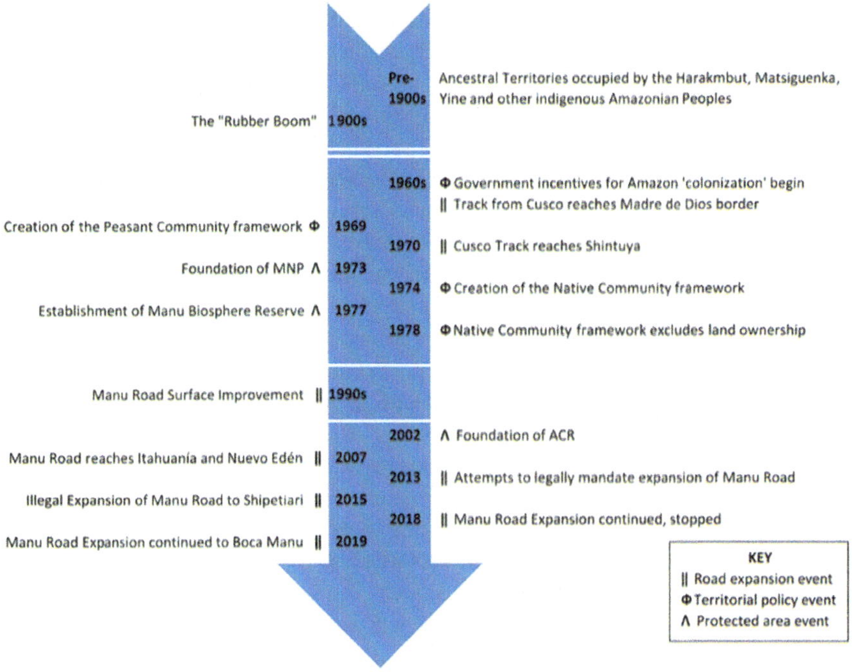

Fig. 2.1 Manu road timeline

In the 1960s, the Peruvian government started efforts to 'colonize' the Amazon by providing various incentives for people to migrate into inaccessible Amazonian areas like Madre de Dios (Gallice et al. 2017; Greene 2009; Harvey and Knox 2015; Larsen 2016; Morel 2014; Shepard et al. 2010). The facilitation of land titling and the expansion of road networks motivated loggers, hunters, farmers and estate managers, 'pioneers' from Andean regions and a few European migrants, to settle in 'the jungle'.[2] This was part of president[3] Fernando Belaúnde's efforts to allow Peruvians to 'conquer their own territory', a strategy that sought to increase political and economic integration, as well as control over natural resources (Greene 2009; Harvey and Knox 2015; Larsen 2016; Morel 2014; Shepard et al. 2010). In Manu, the beginning of this process of 'colonization' was followed by three territorial transformations.

The first was the penetration of the national road network. By the 1960s, the trails that drove trade from Paucartambo, in the Andes of Cusco, to the Kosñipata valley for centuries had been replaced by a modern, although narrow and unpaved, road (Gallice et al. 2017). The military government, which had taken power in 1968, sought to continue these efforts in the early 1970s through the construction of a bridge across the Pillcopata river, a tributary of the Madre de Dios (Gallice et al. 2017; Morel 2014). This expansion continued with the construction of the earliest section the

[2] *La selva*.

[3] His first mandate lasted from 1963 to 1968.

Manu Road itself (Gallice et al. 2017). This 60 km. track connected the settlement of Atalaya, on the Kosñipata province, the Salvación Estate and the Dominican mission of San Miguel de Shintuya (see Map 2.1), around which gathered a population of Indigenous people from various ethnic origins (Gallice et al. 2017; Shepard et al. 2010). Increased access through this road facilitated the entry of Andean migrants, mainly loggers in search of hardwoods and hunters searching for the pelts of giant river otters, jaguars and other wildlife, who formed communities such as Palotoa Llactapampa (Gallice et al. 2017; INRENA 2008).

The second territorial change in Manu was the creation of a protected area. In 1973, in an effort to preserve natural resources and promote tourism in the Manu river basin, motivated by a growing conservation movement, the military government established the MNP, Peru's first national park (Gallice et al. 2017; Ráez Luna 2017; SERNANP 2013; Shepard et al. 2010). The Park was designed as if it was an area free of modern human intervention (Ráez Luna 2017; Shepard et al. 2010). However, various Matsiguenka and Yine communities had lived within the area designated for protection since the rubber 'boom' (Ráez Luna 2017; Shepard et al. 2010). These groups had become accustomed to modern tools and lifestyles facilitated by the protestant missionaries, loggers and hunters who had cohabited with them for decades (Shepard et al. 2010). The Park's regulations allowed these Indigenous groups to stay but restricted their activities to what was seen as their 'traditional' practices (Ráez Luna 2017; Shepard et al. 2010). These restrictions meant that, among other things, these groups had to give up their rifles and go back to hunting with bows and arrows, they had to get rid of their livestock and give up Western medicine, which helped them treat the new illnesses that affected them after contact with Western missionaries and scientists (Shepard et al. 2010). This motivated some Indigenous Yine and Matsiguenka groups to migrate out of the protected area and settle near the mouth of the Manu river, southeast of the Park, later forming communities like Diamante, Isla de los Valles and Shipetiari (Shepard et al. 2010). Evidently, these groups wanted to remain connected to the extractive economies they had already been participating in since they established ongoing contact with outsiders early in the twentieth century (Ráez Luna 2017; Shepard et al. 2010). This area, through which the AMD river and the early section of the Manu Road went through, was declared the Park's buffer zone, and allowed restricted resource use, as well as the settlement of new communities (Gallice et al. 2017). In 1977, UNESCO designated the MNP and its buffer zone as a Biosphere Reserve as part of its strategies to promote sustainable development (Gallice et al. 2017).

The third territorial shift was an attempt at the country-wide recognition of Indigenous territories. Starting in 1969, Peru's military government had been leading a process of agrarian reform which forcefully changed large-scale, estate-based, agriculture for the 'Peasant Community'[4] model, based on the government's understanding of Andean communal organization and small-scale agriculture (Greene 2009; Shepard et al. 2010). Coinciding with the emergence of the Amazonian Indigenous

[4] *Comunidad Campesina.*

movement, the year 1974 saw the creation of the 'Native Community'[5] framework, which sought to allow territorial recognition for Indigenous Amazonian communities (Greene 2009; Larsen 2016; Morel 2014; Shepard et al. 2010). However, it did so by replicating the 'Peasant Community' model, based on the agricultural economy of Andean peoples, instead of recognizing the hunting, gathering and swidden agriculture lifestyle that most Amazonian peoples lived by (Greene 2009; Larsen 2016; Ohl et al. 2008; Peluso and Alexiades 2005; Salisbury et al. 2011; Shepard et al. 2010). In 1978, when the socialist military government was in decline and the country sought to make its natural resources available to the global market once again, the 'Native Community' framework was modified (Larsen 2016). Territorial ownership by Indigenous Amazonian communities was reduced to a government concession to use the land, to which the state kept the ultimate right to (Leal et al. 2015).

The interaction of these territorial changes produced conflicting effects. As is common in frontier contexts, the increase of 'protective' measures motivated the increase of extraction in the unprotected areas (INRENA 2008; Tsing 2005). In the decades after the establishment of the MNP, the increased access provided by the Manu Road stimulated the appearance of Andean logging settlements along its path and even further down the AMD river, like Itahuanía (INRENA 2008). Loggers who had been exploiting large forestry concessions notoriously resisted new timber extraction regulations implemented in the 1990s (INRENA 2008; Larsen 2016; Leal et al. 2015; Salisbury et al. 2011). They continued with their activities in an illegal but very open manner, which evidenced their preference for the state of legal uncertainty that the frontier tends to foster (Greene 2009; INRENA 2008; Shepard et al. 2010; Tsing 2009). In that same decade, the government of Alberto Fujimori sought to increase control of areas vulnerable to guerrillas and cocaine trafficking (Morel 2014). This motivated his government to improve the road from Cusco to Shintuya, which, according to local sources, reduced travelling time from five days to just one. Increased access also prompted the appearance of more Andean settlements like Nuevo Edén and an influx of Matsiguenkas from the rainforests of Cusco to Shipetiari (Gallice et al. 2017; INRENA 2008). All of these communities initially based their economies on timber extraction, which began to have a significant effect on the forest surrounding the MNP (Gallice et al. 2017; INRENA 2008).

Having described the process of increasing migration and resource extraction in Manu, the establishment of protected areas and Native Communities and the expansion of the Manu Road in the midst of these events, we now turn to the most recent events regarding this infrastructure. The following section provides details about the controversy generated by the recent attempts to expand the Manu Road towards Boca Manu, a settlement considered to be the 'entrance' to the MNP (Gallice et al. 2017).

[5] *Comunidad Nativa.*

2.2 Recent History of the Manu Road, the Clash of Conservation and Modernization

In this section, we turn to the latest events regarding the expansion of the Manu Road. The dispute between the conservation sector and local elites and communities regarding the future of the area is quite evident in this period. The political confrontation surrounding this attempt to expand the road is key to understand our discussion in further chapters, as it begins to show the influence of extra-local powers in this process.

By 2002, because of the ever increasing pressure from logging and colonization in the area, multi-ethnic Indigenous organizations concluded the process to establish the ACR (INRENA 2008). The creation of the reserve sought to set aside the use of traditional territory for Harakmbut, Matsiguenka and Yine Native Communities but it later seemed evident that participatory measures did not grant them the necessary influence to do this at will (INRENA 2008; Oliart and Biffi 2010). With the establishment of the ACR, both sides of the AMD river, and the Manu Road so far, became protected areas. Nevertheless, between 2000 and 2007, the Manu Road was extended once more after a long hiatus, connecting the Andean settlements of Itahuanía and Nuevo Edén to the city of Cusco (Gallice et al. 2017).

On that same year, the Madre de Dios Regional Government (GOREMAD) also submitted a proposal to improve an, allegedly, already existing footpath that connected farming plots between Nuevo Edén to the area across the AMD river from the settlement at Shipetiari (Gallice et al. 2017). By then, the construction of the Interoceanic Highway,[6] which would cross the Andes and the southern margins of the Madre de Dios region to connect the Peruvian ports of the Pacific with Brazil's road network, had begun generating much anticipation (Harvey and Knox 2015; Pieck 2013, 2015; Zibechi 2006). Thus, GOREMAD continued its efforts to expand the Manu Road. In 2009, it presented a second proposal, this time to expand the Manu Road from its end point to Boca Manu, considered to be the entrance to the MNP, and further to Boca Colorado, a mining town which would soon be connected to the Interoceanic Highway (Gallice et al. 2017). This attempt to expand the road was rejected by the Ministry of Economics and Finance (MEF), in charge of assigning budgets to regional governments, given the lack of the required technical assessments (Gallice et al. 2017).

In 2012, representatives of the Madre de Dios region at the National Congress persisted by using a different strategy (Congreso de la República 2013; Gallice et al. 2017). They proposed a law[7] that sought to declare the construction of a 'Regional Integration Road' from Nuevo Edén through Boca Manu and to Boca Colorado (see Map 2.1) as a 'public necessity' (Congreso de la República 2013, p. 1; Gallice et al. 2017). Being declared as such would allow the prioritization of this project and, thus, the avoidance of the red tape which had legally vetoed the previous attempt

[6] IIRSA Sur or *Carretera Interoceánica*.

[7] Proyecto de Ley No. 2320/2012-CR.

(AIDESEP 2018; Congreso de la República 2013; Ráez Luna 2017). The proposal was approved for debate by congress's Transport and Communications Committee but was opposed by the MEF (Congreso de la República 2013; MINAM 2016). Technically, such a declaration could not be made by the congressional approval of a law but could only be declared by a competent Ministry (Congreso de la República 2013; MINAM 2016). Furthermore, it did not follow the national strategy of multi-modal transport, which seeks to promote alternative means of transportation in sensitive areas by connecting road, river and aerial transport routes (Barrantes et al. 2014). Thus, the construction of the Manu Road began to turn into an intra-state conflict, where the representatives of the GOREMAD would confront several bodies of the Peruvian National Government (Bryant and Bailey 2000).

The following year, the Manu Road became an important part of Luis Otsuka's campaign to become Madre de Dios's Regional Governor. As a prominent miner and leader of the regional mining federation, Otsuka's campaign was based on his vision of natural resource extraction as the only means for the development of the region (Torres López 2016; Walker 2012). After being elected as Regional Governor, he rekindled the debate about the road project and faced the opposing reaction of the National Parks Service[8] (SERNANP), which had not given its legally required approval for the road (MINAM 2016). Despite not having approval from MEF, SERNANP and other National Government bodies, Otsuka's GOREMAD started building the road in June 2015 (Gallice et al. 2017; Torres López 2016). According to Ricardo Gómez, manager of the Manu Sub-Regional Government (SRG), and Gallice et al.'s research (2017), the building of this stretch of road was supported by the previous proposal to improve the footpath between Nuevo Edén and Shipetiari, turning it into a 'local track'.[9] The new stretch of road, however, went beyond the access point to the settlement at Shipetiari onto a dead end further away and closer to the territory of the Native Community of Diamante (see Map 2.1) (Gallice et al. 2017; Torres López 2016). The width of the road also exceeded that of a footpath or a 'local track', being enough for heavy vehicle traffic (Gallice et al. 2017; Larrea-Gallegos et al. 2016). Consequentially, following an inspection, a judge from Cusco sued Otsuka, Gómez and several of the engineers in charge of the physical works for illegal deforestation (Gallice et al., 2017; Torres López, 2016). This legal action prompted the SRG to interrupt the works (see Map 2.1) (Gallice et al. 2017; Torres López 2016).

This interruption and the legal actions taken against the road's most prominent advocates were the motivation for two protests in late 2015. In September, community members from Diamante, Boca Manu and Isla de los Valles gathered in Boca Manu and detained all boats attempting to enter or leave the MNP, which got the attention of international media (Gallice et al. 2017; Post 2015; Torres López 2016). In November, a large number of people from various communities in the Manu province arrived in the provincial capital of Salvación to protest in favour of the road (witnessed personally). This coincided with a regional protest, led by Otsuka

[8] *Servicio Nacional de Áreas Protegidas por el Estado.*

[9] *Camino Vecinal.*

himself, in which miners opposed new national regulations that sought to control the impact of their illegal trade (Post 2015; Torres López 2016; Walker 2012). Protesters in Manu blocked the access routes to the provincial capital and made MNP rangers march barefoot in front of their picket, a powerful symbolic gesture (witnessed personally). Both protests ended through the intervention of the SRG. Given the lack of results from the two protests, and of any immediate efforts by the SRG to continue building the road, in early 2016, off-road communities joined forces to continue building the road themselves in a series of communal working days[10] (Torres López 2016). They used their own machetes and chainsaws in an attempt to clear a stretch of forest, delineated on their own criteria, from the settlement in Diamante to the end of the road constructed by the SRG.[11] Despite having gathered support from local merchants and recruiting several women as cooks, the lack of food and strenuous working conditions prevented the troop from finishing the job. The lack of heavy machinery prevented proper treatment of the surface and the cleared stretch was overgrown by the forest soon after (local sources). Communal leaders from Diamante portray this failed effort as another way of protesting in favour of the road. This type of action, even through failure, is an integral part of struggles for road construction in Peru (Harvey and Knox 2012, 2015).

By mid-2017, while Eduardo conducted fieldwork in the area, the SRG's Ricardo Gómez was seeking to build the road 'formally', to secure the required budget and authorizations. Regular meetings between communal authorities and GOREMAD representatives had been taking place since May 2016 to discuss the progress of the project and ensure communal collaboration in the formal proceedings (MINCUL 2016a–c). Despite the initial opposition from the Peruvian National Government through various Ministries and SERNANP, the political context changed in favour of the proponents of the road after the presidential transition of July 2016. The government of nationalist Ollanta Humala was in office until that month, after which Peru's neoliberal party, led by Pedro Pablo Kuczynski, took office. Gómez himself said that 'the difference in SERNANP's attitude can be seen from far away', indicating a more favourable outlook towards the road from Peru's new National Government.

Resulting from this political change was SERNANP's April 2017 letter, which accepted the road project as 'compatible' with the goals of the surrounding protected areas (SERNANP 2017, p. 1). According to Rómulo Oropesa, an MNP ranger, SERNANP had relented to the pressure from the communities and was seeking to adapt MNP's upcoming master plan to the new situation presented by the road. The previous strategy to resist road construction had been changed towards adapting the MNP's upcoming master plan to the new situation presented by the road by seeking resources to establish new guard stations and increasing the number of rangers. The even more under-resourced and threatened ACR would possibly attempt something similar. At the time of publication, the only stance taken by the management of either

[10]'*Faenas*'.

[11]This action and its effects are not clearly recognized by the works of Gallice et al. and Larrea-Gallegos et al., who base their observations on satellite imagery produced by Finer for the MAAP Project (Finer et al 2016; Gallice et al. 2017; Larrea-Gallegos et al. 2016).

protected area is a brief proposition to increase vigilance and improve participation measures of the ACR, without any details regarding the way this would be done (SERNANP 2016).

Ricardo Gómez and Neptalí Gutierrez, former Chief of Diamante and self-appointed 'President of the Road', worked throughout 2017 and 2018, and finally received approval to build the road in October 2018[12] (John and Munro 2019). The ceremony to inaugurate the works was conducted in mid-November 2018, in a rush to finish building it before the end of Luis Otsuka's GOREMAD term in January 2019 and before the beginning of the rainy season (John and Munro 2019). These attempts at taking credit for building infrastructure before work is finished on the following term are usual in Peru (Harvey and Knox 2015). Despite the efforts, the weather conditions did not allow for the completion of these works and the road hadn't reached Diamante by the end of 2018 (John and Munro 2019). By early June 2019, the new GOREMAD authorities were only waiting for paperwork to be completed and had determined that the works to complete the road all the way to Boca Manu would start later that month (COHARYIMA 2019). This was done with urgency given the 'worry and displeasure"'of the off-road Native Communities regarding the lack of completion of the road (COHARYIMA 2019).

In this chapter, we have provided a summary of the rich history in Manu, mainly in the last 50 years, and the recent events regarding the expansion of the Manu Road. Events like the establishment of the MNP and the ACR, as well as that of the Native Community framework and the gradual expansion of the Manu Road (Gallice et al. 2017), are the foundations to aspects discussed in further chapters. Similarly crucial is the political contention surrounding the expansion of the road. The confrontation between GOREMAD and protected area management authorities affiliated to the National Government reveals the tensions between the conservation sector and regional authorities pushing for modernization, which are at the root of contradictions further explored in the following chapters. But before that, in the next chapter, we introduce key debates in the broader literature that will help us unpack the multilayered frictions existing in Manu.

References

AIDESEP (2018) Informe sobre la situación de los pueblos en aislamiento y contacto inicial de la Amazonía peruana. Lima, AIDESEP

Ames P (2010) Desigualdad y Territorio en el Perú: una geografía jerarquizada. Argumentos 1

Barrantes R, Fiestas J, Hopkins Á (2014) Evolución de la infraestructura de Transporte y Energía en la Amazonía Peruana. In: Barrantes R, Glave M (eds) Amazonía Peruana y Desarrollo Económico. GRADE, IEP, Lima, pp 109–160

Bryant R, Bailey S (2000) Third world political ecology. Routledge, London and New York

[12] Eduardo also followed this process through the Facebook sites of the Manu SRG and GOREMAD, and through contact with local sources.

COHARYIMA (2019) Acuerdan Reiniciar Obras de Camino Vecinal Maronal—Boca Manu. http://www.coharyima.org/noticias/acuerdan-reiniciar-obras-de-camino-vecinal-maronal-boca-manu?fbclid=IwAR1ydWz6I1Yu1sz59PcmAJQwtHJzWCmdMaTgvpysV0zZZB8ntak09tBP334

Congreso de la República (2013) Proyecto de ley que Declara de Necesidad Pública e Interés Nacional la Construcción de la Carretera de Integración Regional de Madre de Dios. Comisión de Transportes y Comunicaciones, Lim

Finer M, Novoa S, Olexy T (2016) Construction of a new highway between Manu National Park and Amarakaeri Communal Reserve (Madre de Dios). maaproject.org

Gallice G, Larrea-Gallegos G, Vásquez-Rowe I (2017) The threat of road expansion in the Peruvian Amazon. Oryx 1–9

Greene S (2009) Caminos y Carretera: Acostumbrando la indigenidad en la selva peruana. IEP Instituto de Estudios Peruanos, Lima

Harvey P, Knox H (2012) The enchantments of infrastructure. Mobilities 7(4):521–536

Harvey P, Knox H (2015) Roads: an anthropology of infrastructure and expertise. Expertise, Ithaca, NY

INRENA (2008) Plan Maestro de la Reserva Comunal Amarakaeri 2008–2012. Ministerio de Agricultura, Lima

John B, Munro E (2019) Voices on an Amazon road—Report to Funders. Scientific Exploration Society, UK.

Larrea-Gallegos G, Vazquez-Rowe I, Gallice G (2016) Life cycle assessment of the construction of an unpaved road in an undisturbed tropical rainforest area in the vicinity of Manu National Park, Peru. Int J Life Cycle Assess 11(17/2016):1–16

Larsen PB (2016) Derechos indígenas, gobernanza ambiental y recursos en la Amazonía peruana, Hacia una antropología de la posfrontera. Instituto de Estudios Peruanos, Lima

Leal D, Salisbury D, Faquín Fernandez J, Cauper Pezo L, Silva J (2015) Ideas cambiantes sobre territorio, recursos y redes políticas en la Amazonía indígena: un estudio de caso sobre Perú. J Latin Am Geogr 14(2):181–204

MINAM (2016) Dossier: ¿Por qué no debe aprobarse el predictamen de insistencia de la ley sobre la carretera Madre de Dios, tramo Nuevo Edén—Boca Manu—Boca Colorado? Ministerio del Ambiente, Lima

MINCUL (2016a) Minutas 3 Reunión Carretera 17 05 16. Not published

MINCUL (2016b) Minutas 4 Reunión Carretera 21 07 16. Not published

MINCUL (2016c) Minutas 5 Reunión Carretera 04 08 16. Not published

Morel J (2014) De una a muchas Amazonías: Los Discursos sobre 'La Selva' (1963-2012). In: Barrantes R, Glave M (eds) Amazonía Peruana y Desarrollo Económico. GRADE, IEP, Lima, pp 21–46

Ohl J, Wezel A, Shepard GH, Yu DW (2008) Swidden agriculture in a protected area: the Matsigenka native communities of Manu National Park, Peru. Environ Dev Sustain 10(6):827–843

Oliart P, Biffi V (2010) Territorialidad indígena, conservación y desarrollo: Discursos sobre la biodiversidad en la Amazonía peruana. Instituto del Bien Común, Lima

Peluso DM, Alexiades MN (2005) Indigenous urbanization and Amazonia's post-traditional environmental economy. Trad Dwell Settl Rev 16(2):7–16

Pieck SK (2013) Asphalt dreams: road construction and environmental citizenship in Peru. Dev Change 44(5):1039–1063

Pieck SK (2015) "To be led differently": Neoliberalism, road construction, and NGO counter-conducts in Peru. Geoforum 64:304–313

Post C (2015) Peru state in standoff with government over illegal mining. https://perureports.com/peru-state-in-standoff-against-government-over-illegal-mining/2728/

Ráez Luna E (2017) Pueblos Indígenas y Conservación de la Naturaleza: ¿Debates teóricos o urgencias vitales? El caso del Parque Nacional del Manu. In: Ráez Luna E (ed) Derechos Ancestrales y Conservación de la Naturaleza en Debate: El caso de los Pueblos Indígenas del Parque Nacional del Manu. Seminario Permanente de Investigación Agraria (SEPIA), Lima

Salisbury DS, López JB, Vela Alvarado JW (2011) Transboundary political ecology in Amazonia: history, culture, and conflicts of the borderland Asháninka. Journal of Cultural Geography 28(1):147–177

SERNANP (2013) Parque Nacional del Manu, 40 años (SERNANP). Biblios, Lima

SERNANP (2016) Plan Maestro de la Reserva Comunal Amarakaeri 2016–2020. Ministerio del Ambiente, Lima

SERNANP (2017) Oficio No 620–2017—SERNANP—DGANP. Ministerio del Ambiente, Lima

Shepard GH, Rummenhoeller K, Ohl-Schacherer J, Yu DW (2010) Trouble in paradise: indigenous populations, anthropological policies, and biodiversity conservation in Manu National Park, Peru. J Sustain Forest 29(2–4):252–301

Torres López F (2016) The highway tearing through the heart of the Amazon. Ojo Público 1–22. http://ojo-publico.com/sites/apps/amarakaeri-parte1-english/

Tsing AL (2005) Friction: an ethnography of global connection. Princeton University Press, New Jersey

Tsing AL (2009) How to make resources in order to destroy them (and then save them?) on the Salvage Frontier. In: Histories of the future, pp. 51–73

Walker R (2012) Peru' s gold rush pits illegal miners against government. https://www.bbc.com/news/world-latin-america-18524330

Zibechi R (2006) IIRSA: la integración a la medida de los mercados. Ecología Política 31:19–25

Chapter 3
Uncovering the 'Enchantments of Infrastructure', Territory and Power Relations in Context

Abstract This chapter explores some key concepts and frameworks that provide analytical lenses to unpack the complex realities of the Manu Road. Alongside Harvey and Knox's concept of 'enchantments of infrastructure' (Mobilities, 7:521–536, 2012), we explore past and present processes of colonisation and modernization, as well as the particularities of postfrontier spaces, territorialisation and power relations at play in this region. Using these theoretical lenses to disentangle the intricacies of the tensions around the Manu Road from different angles not only facilitates a deeper understanding of its 'enchantments', but also provides a crucial foundation for the analysis of ethnographic material in the following chapters.

Keywords Manu road · First roads · Frontier infrastructure · Enchantments · Indigenous · Amazonia · Political ecology · Internal colonization · Postfrontier · Colonisation · Modernisation

3.1 The Incomplete 'Colonization' of the Peruvian Amazon

Having examined the historical background of the case of the Manu Road in Chap. 2, it is important now to look at some broader theoretical considerations that will help unpack this case and its implications. But these considerations, as well as most of the key issues we raise along with this book, are not only relevant to this particular road. In fact, the tensions and processes at play around the Manu road are also present in the rest of Peru and neighbouring countries. Just in the Peruvian Amazon, several road construction projects have been in the making in all its regions in recent years. For instance, the Madre de Dios region has had two congressional proposals to expand its road network in ways that would affect protected areas and Indigenous territories (Congreso de la República 2013, 2017). The Ucayali region of the central Peruvian Amazon has also had its own wave of road expansion projects, including one for a road that cuts through territories reserved for isolated Indigenous groups (SPDA 2017a, 2018). The Loreto region, in the north, has also seen similar road projects (El Peruano 2017; SPDA 2017b) and a larger initiative for the construction of a regional waterway (Portugal and Linares 2018), both with significant social and environmental consequences. Warnings from the environmental sector and the Indigenous rights

E. Salazar Moreira and M. Palomino-Schalscha, *Road Expansion in the Peruvian Amazon*, SpringerBriefs in Latin American Studies, https://doi.org/10.1007/978-3-030-47182-8_3

movement about these projects have not had any significant consequence (AIDESEP 2018b; Bonifaz 2018; SPDA 2017a). In this context, the extension of the Manu Road can be considered an especially relevant case due to its pivotal place with both conservation and development. It is not only likely to affect Peru's first and most emblematic protected area (Gallice et al. 2017), but is also one of the first such projects to have been proposed after the completion of the Interoceanic Highway (Congreso de la República 2013).

In this brief section, we explain key concepts of the 'colonization' of the Peruvian Amazon and situate the reader in its political backdrop (Larsen 2016; Morel 2014). The story of increasing immigration and gradual connectivity into Manu, described in Chap. 2, is not unique to this area, as it followed a model of 'colonization' that was actuated throughout the Peruvian Amazon in the late 1950s and the 1960s (Larsen 2016; Morel 2014). As it had been since the time of Spanish colonial rule, the 'jungle' was mostly seen as a pool of resources, and in this period, the aim was to place these under the state's influence in order to generate economic development (Larsen 2016; Morel 2014; Royo-Villanova 2017). President Fernando Belaúnde's writings about 'The conquest of Peru by Peruvians'[1] epitomized the discourse behind this set of developmental strategies, mostly inspired by those promoted by the Alliance of Progress to curb the spread of communism in the 'Third World' (Larsen 2016; Morel 2014).

Following the developmental concepts of the time, 'The conquest of Peru by Peruvians' was part of a national strategy that sought to 'improve' the country through modernization (McEwan 2009; Morel 2014). The new 'conquest' would contribute to this greater national goal through a regional strategy of territorial expansion that would populate the resource-rich but supposedly uninhabited Amazonian rainforest (Morel 2014). Thus, landless Andean populations were incentivized through facilitated land titling and various other means to migrate east into what was seen by the urban coastal elites that ran the country as the empty and unproductive 'wilderness' of the Amazonian frontier (Barbier 2012; Barrantes et al. 2014; Hecht 2008; Larsen 2016). In this way, the Amazonian regions would be transformed into a profitable frontier territory through resource exploitation and agriculture (Morel 2014; Vadjunec et al. 2011).

Apart from being influenced by the modernist trend of the era, this approach to development was also shaped by the social structures of colonial Peru, which placed Andean Indigenous people as subordinate to European descendants and mixed-race *mestizos* but superior to the 'savage Indians' of the Amazon[2] (Ames 2010; Morel 2014). In this way, the attractiveness of the abundant resources of Amazonian territories would bring more 'civilized' Andean populations that would contribute to the development of the 'savage' Amazonian 'other' through a process of internal

---

[1]'*La conquista del Perú por los peruanos*', was published in 1959 by Belaúnde, who would soon become President (Morel 2014, p. 25).

[2]This conception of Peru as a nation comprised only by coastal elites and the peasants of the Andean highlands, invisibilizing Amazonian peoples, influences politics and scholarship to this day (Greene 2006).

colonialism (Cupples et al. 2019; McEwan 2009; Mignolo 2000; Vadjunec et al. 2011). It was thus assumed that the implementation of this strategy would contribute to the economic growth of Peru by increasing state control over Amazonian natural resources and peoples, and by alleviating the crisis of land availability in the Andes (Barrantes et al. 2014; Larsen 2016; Morel 2014).

Here it becomes useful for us to define what is understood by 'the frontier', a term that captures much of the process just described and its motivations. In doing this, it is necessary to recognize the existence of what Harvey and Knox call the 'myth of the frontier', which essentializes these spaces for outsiders to appropriate them (2015, p. 55). Frontier spaces are purposefully portrayed as empty of the presence of the state and the order it supposedly guarantees, as bereft of modernity and civilization and as deficient and 'wild' (Barbier 2012; Harvey and Knox 2015; Larsen 2016; Murray-Li 2014; Tsing 2005; Watts 2018). Thus, as spaces in the 'wrong' side of certain binaries, they are seen as spaces in need of 'improvement' and outside intervention. At the same time, frontier spaces are represented as full of generative potential given the abundance of land, resources and opportunities that are freely available to whoever is willing to venture into these seemingly unruly zones. The combination of these paradoxical notions justifies and motivates the intervention of outsiders in order to realize the potential of the frontier, which gives way to conflict and violence for the sake of their increased control of the resources present in these spaces. The proliferation of this narrative by agents that seek to benefit from the access it provides to them invisibilizes people that already inhabit these spaces and threatens their rights, knowledge and places (Larsen 2016; Murray-Li 2014; Tsing 2005). Their invisibilization and displacement, then, leads to the uninhibited exploitation of the resources present in the spaces they inhabited by incoming outsiders (Barbier 2012; Larsen 2016; Murray-Li 2014; Tsing 2005; Watts 2018). Through the recognition of this essentialist frontier narrative, the frontier can be defined as 'a social space within which forms of rule and authority, and multiple sovereignties, are in question' (Larsen 2016; Watts 2018, p. 480). Thus, the limits of the order of the state in frontier spaces do not mean that there is no order at all, and the absence of 'modernity' or 'development' does not mean that there is no wellbeing, as those who originally inhabited them had their own forms of order and sources of wellbeing (Watts 2018).

The following step in the process of 'colonization' of Peru's Amazonian frontier by Andean migrants and their establishment in productive communities would be to connect them with Andean urban areas and their markets through a network of roads (Morel 2014). These roads would also be the means through which the state would enforce land sovereignty over its Amazonian territories and resources (Barrantes et al. 2014). This step in the process of Amazonian modernization also considered the exploitation of natural resources, including their depletion, and the degradation of forests, as a justified cost of development, a common occurrence in frontier-making processes (Pieck 2011; Tsing 2005). Through the proposed road network, the apparently limitless extractive potential of the Amazonian rainforest would be made available to the international demand for commodities (Jiménez 2017; Vadjunec et al. 2011). Moreover, Amazon's supposed aptitude for agriculture was expected to turn this territory into a 'national pantry' that would serve urban markets throughout

Peru via the proposed roads (Morel 2014). In all, road connectivity would allow the fulfilment of the forest's promise to the country's economy and its global clientele, and of the country's promise of development to those who lived in the forest (Harvey and Knox 2015).

Considering the process just described will facilitate the understanding of its consequences in the Madre de Dios region and the Manu province, and how its lack of completion in these areas influences the current aspirations for road expansion. This process also has implications in the power relations and colonial structures discussed further on. We turn to consider the generation of aspirations for modernization in the next section.

3.2 Incomplete Connectivity and Aspirations of Modernization

This section explains the way in which the Madre de Dios region was not fully incorporated into 'the conquest of Peru by Peruvians' project and how this entrenched aspirations for modernization in both Andean migrants and Indigenous Amazonians in the region (Morel 2014). This is a key element to understand the capacity of roads to 'enchant', which is explored in the following section (Harvey and Knox 2012).

The process of Amazonian colonization promoted by Belaúnde, however, was never fully implemented in the Madre de Dios region given a change in the national government's priorities after the military coup of 1968 (Barrantes et al. 2014). The governments after Belaúnde's focused on oil extraction in the northern Peruvian Amazon and the roads that would connect the far-reaching communities of Madre de Dios were either built precariously or not built at all (Barrantes et al. 2014; Harvey and Knox 2015). This lack of completion of proposed road networks has taken place in the Brazilian context as well, with similar consequences (Araujo et al. 2011; Arima et al. 2005, 2013; Carvalho et al. 2002). Connectivity with the rest of Peruvian Amazonia through the Amazon river system, which would have been an alternative to road connectivity, was not possible either as the Madre de Dios watershed only joins that of the Amazon river far into Brazilian territory (Figallo and Vergara 2014). Furthermore, the poor soils of the Amazonian rainforest did not allow for the promised agricultural productivity, and the isolated Andean migrant communities were left with the exploitation of timber as their only viable livelihood (Figallo and Vergara 2014). Thus, the further stages of frontier development were never reached and the economic survival of these isolated migrant communities became tied to the boom and bust cycles that the precarious exploitation of natural resources tends to foster (Barbier 2012).

The isolation that these communities of Andean settlers were left in after migrating to the frontier and not being connected to the road network as expected are very likely to have generated perceptions of abandonment and seclusion (Barrantes et al. 2014; Harvey and Knox 2012; Larsen 2016; Morel 2014; Vadjunec et al. 2011;

Wilson 2004). These perceptions, together with widespread discourses around modernization, tend to turn the expectations of development that frontier dwellers hold into overwhelming aspirations (Harvey and Knox 2008, 2012; Uribe 2017). Thus, the mere anticipation generated by forthcoming roads still motivates reactions based on the imagination of what these infrastructures could bring (Campbell 2012; Haines 2017; Harvey and Knox 2012, 2015). Not surprisingly, this perception of isolation remains as the focus of the continued promises of infrastructural development that the state and other powerful agents direct at these communities (Uribe 2017).

The aspirations of modernization held by isolated frontier dwellers are coupled with the cosmopolitan ambitions shared by many rural communities in the global south (Harvey and Knox 2012; Tsing 2005). The evocative power that the discourse of modernization holds seems to put the possibility of globalized consumption in much closer reach (Tsing 2000, 2005; Uribe 2017). This cosmopolitanism, perceived as something that 'should be' through the power of development, appears to be more easily achievable by the impending construction of roads (McEwan 2009; Tsing 2005; Uribe 2017). Therefore, these infrastructures would seemingly facilitate breaching through the 'traditional versus modern' binary (Uribe 2017). The aspirations to join the universal, the global, and escape the enclosure of isolation that Andean migrants and other frontier dwellers have, thus, pave the way for an unrelenting desire for connectivity to the national road network (Harvey and Knox 2012, 2015; Tsing 2005).

Furthermore, contrary to essentialist notions based on the idea of the 'noble savage', these aspirations of modernization are often adopted and shared by Indigenous peoples in the frontier (Adams and Hutton 2007; Kohler and Brondizio 2016; Tsing 2005, 2009). Amazonian Indigenous peoples, who form a complex and diverse social landscape, are not the exception, and constantly adapt their outlooks and livelihoods to the new conditions around them (Figallo and Vergara 2014; Leal et al. 2015; Oliart and Biffi 2010; Peluso and Alexiades 2005; Vadjunec et al. 2011). The degradation of the ecosystems that traditional livelihoods depended on is a significant factor of pressure for Indigenous frontier dwellers to adjust their livelihoods (Sunderlin et al. 2005; Tsing 2009). Moreover, the increasing contact of Amazonian Indigenous peoples with settlers, loggers, merchants, government officials, tourists and other incomers, who articulate the Amazonian frontier with the global economy, is another key incentive for this adaptation (Figallo and Vergara 2014; Leal et al. 2015; Oliart and Biffi 2010). Most importantly, changes in preferences, their precarious inclusion into the state's services system and the immense economic interest in the resources their territories hold often generate an unprecedented need to access cash (Oliart and Biffi 2010; Ráez Luna 2017). This need, and the impracticality of satisfying it through 'traditional practices' as required by the restrictions of conservation, becomes a significant incentive to modify their livelihoods in ways that often do not fit said restrictions (Oliart and Biffi 2010; Ráez Luna 2017). In the case of communities within the MNP, Ráez Luna (2017) identifies a process of acculturation between Matsiguenka communities and Andean migrants, as well as other incomers, which implies a mutual yet asymmetric process of cultural change that significantly affects the preferences and practices of the former. The territorial nature of a frontier process led by a dominant elite, as well as the colonial structures inherited through the process of colonization by Andean migrants, underlies this process of cultural adaptation (Escobar 2008; McEwan 2009; Peluso and Lund 2011; Uribe 2017). Given these conditions, it can be said that for different yet intertwined reasons, and often not because of their own choosing, both Indigenous Amazonian peoples and Andean migrants in Manu share aspirations around the promises of roads. These aspirations establish the conditions for roads to 'enchant' (Harvey and Knox 2012). Having covered these underlying aspects, we can now turn to an evaluation of the capacity of roads to 'enchant' more generally, and the nature of their 'enchantments'.

3.3 'Enchanting' Roads

This section explains Harvey and Knox's concept of 'the enchantments of infrastructure', its various elements and its applicability to the case of the Manu Road (2012). In doing so, it provides analytical tools to unpack the aspirations held by isolated frontier dwellers in this area, as well as of the likelihood of these not to be fulfilled and how this, paradoxically, is the foundation of the capacity of roads to 'enchant'. Elements specific to Manu, a province where the interests of conservation and extraction clash, and their influence on the 'enchantments' of the Manu Road are explored in subsequent sections.

In 'The Enchantments of Infrastructure', Harvey and Knox (2012) identify a set of 'promises' that roads make to frontier dwellers with such overwhelming aspirations for modernization. These promises allude to what Tsing calls 'universals', notions based on global imaginaries of development that motivate the desire to adapt to 'internationally mandated standards of progress' (2005, p. 9). Views of what these standards mean have changed in time; at first, roads were seen as a manifestation of state presence and later signified integration to the market, but have maintained connectivity as their main value (Campbell 2012). Despite the anticipation that these promises of modernization generate, and the assumed predictability of roads (Howe et al. 2016), the failure of these infrastructures in fulfilling the type of development they promise is highly likely and, in fact, to be expected (Harvey et al. 2017; Harvey and Knox 2012; Larkin 2013). Moreover, Larkin analyses the way in which the promises of roads often exceed their capacity but are taken at face value nonetheless, concluding that their popularity is based on what he calls 'infrastructural fetishism' (2013). The universal notions that evoke this 'fetishism' face what Tsing dubs as 'friction' when they encounter the specific situations that are presented in local contexts, which are far from consistent with global imaginaries (2005). In fact, what seems to many as the natural process of development is based on Western notions of modernization imposed in the global south and leading to a number of disparate outcomes (McEwan 2009). Therefore, it is the very nature of the frontier, where sovereignty over newly accessible territories is contested and the outcomes of power relations are uncertain, which prevents roads from generating the changes that these universal notions seem to anticipate (Harvey and Knox 2012, 2015). Once built, these fragile and unstable Amazonian roads still present an astonishingly difficult layout, and even introduce new unexpected risks (Howe et al. 2016; Uribe 2017). This is especially so when frontier roads are built without planning about their use or considerations additional to their sole construction, which is often the case in the Peruvian Amazon (MINAM 2016; Vergara et al. 2014).

However, according to Harvey and Knox, it is the very lack of fulfilment of these infrastructural 'promises' that perpetuates the aspirations of development that frontier dwellers hold, thus leading to 'enchantment' (2012). The significant gaps between the abstract intentions of modernizing infrastructure projects and the actual effects that built infrastructures produce maintain these aspirations unsatisfied and perpetuate them (Harvey and Knox 2012; Tsing 2005). More so, the purposeful exclusion of the frontier by the state preserves various forms of violence that later favour what modernizing projects that serve extraction promise to frontier dwellers (Uribe 2017). It seems that the very failure of development projects generates incentives to insist in their application (Murray-Li 2007).

Considering the nature of these 'enchanting promises' of modernization illustrates this point further (Harvey and Knox 2012). The first of these 'promises' is the perception of the increased speed that roads provide for transportation, which is expected to end the 'tyranny of delay' that affects off-road livelihoods through safer, faster and more dignified travel (Campbell 2010; Harvey and Knox 2012, p. 524; Tsing 2005; Uribe 2017). The speed provided by roads is also the basis for other elements of development, facilitating the transport of services and goods (Harvey and Knox 2012; Pieck 2011; Wilson 2004). Nonetheless, frontier roads often lack enough surface quality and are too fragile to deliver enough speed (Harvey et al. 2017; Kernaghan 2012). This effect is intensified also by their usual lack of maintenance, something that has been observed in Peru particularly (Barrantes et al. 2014; Harvey et al. 2017; Kernaghan 2012). The lack of quality of frontier roads perpetuates aspirations to have more and better roads, feeding their power to 'enchant' (Harvey and Knox 2012; Uribe 2017). Furthermore, and regardless of their surface quality, frontier roads are not only used by outgoing frontier dwellers. These frontier roads also invite people from other regions to immigrate, increasing the potential for various sorts of conflicts, which often take place in the Peruvian Amazon (Barbier 2012; Barrantes et al. 2014; Gallice et al. 2017; Leal et al. 2015; Pieck 2013; Tsing 2005, 2009; Uribe 2017; Vadjunec et al. 2011). For example, the migration facilitated by road access increases the possibilities of extreme land use change and depletion of resources within Indigenous territories, which motivates Indigenous actors to perceive Andean migrants as aggressive incomers (Barrantes and Glave 2014; Leal et al. 2015; Oliart and Biffi 2010; Salisbury et al. 2011).

The second change that roads promise to deliver is an increased integration with previously inaccessible political centres (Harvey and Knox 2012). This is expected to satisfy the aspirations of increased political power that off-road communities hold, ending the feelings of abandonment and disconnection that their members often have (Barrantes et al. 2014; Campbell 2010; Harvey and Knox 2012; Morel 2014;

Sunderlin et al. 2005; Uribe 2017). Through political integration, roads are also predicted to allow the manifestation of state presence, increase access to government development strategies and give communities the ability to make their governments accountable (Campbell 2010; Harvey and Knox 2012, 2015; Uribe 2017). As a consequence, being integrated to political centres is also expected to grant access to public services like electricity, communication technologies, water and sanitation and to improve education and healthcare (Harvey and Knox 2012; INRENA 2008). Despite this aspiration, the entirety of the Amazonian territories of Peru has remained a low priority in the expansion of state-provided services, especially of energy provision (Barrantes et al. 2014). And, even though increased state presence and law enforcement are aspired to, roads can also generate rejection of law enforcement from frontier dwellers who contest state authority and foster corruption in the absence of adequate supervision of law enforcement agents (Harvey and Knox 2008; Sunderlin et al. 2005; Wilson 2004). Moreover, usually the haphazard way in which roads in the Peruvian Amazon are built can itself be a manifestation of state absence, which in turn perpetuates the 'enchantment' of political integration (Harvey and Knox 2012, 2015; Kernaghan 2012).

The third, and possibly the most anticipated, promise that these infrastructures make to frontier dwellers is the connection with regional and national markets (Harvey and Knox 2012, 2015). Roads are anticipated to provide previously off-road communities access to trade and new sources of income, and give outsiders access to the resources available in the frontier, generating economic growth through market relations (Campbell 2010; Harvey and Knox 2012, 2015; Pieck 2011). More efficient connectivity with the market also forebodes an increased access to cash, which can be expected to become especially appealing to Indigenous Amazonian people (Oliart and Biffi 2010; Ráez Luna 2017).

However, in the specific case of Amazonia, the expected sources of increased income are not as profitable as they seem (Adams et al. 2004; Barbier 2012; Figallo and Vergara 2014; Leal et al. 2015). In the case of Manu, the main economic activity is logging, followed by banana and yucca farming, which both Indigenous Amazonians and Andean migrants incorporate in different ways (INRENA 2008). The low quality of Amazonian soil limits the productivity of agriculture, and the perishable nature of products like bananas and yuccas affects their transportation, which makes farming a meagrely lucrative livelihood at best (Figallo and Vergara 2014; Sunderlin et al. 2005). Furthermore, the unfavourable terms in which frontier farmers compete in the wider market can engender abusive trade relations that increase poverty and inequality, something that is not exclusive to the Peruvian Amazon or to farming (Adams and Hutton 2007; Murray-Li 2014; Sunderlin et al. 2005; Wilson 2004). As is usual in Peruvian Amazonia, logging is conducted in an illegal and unmanaged manner that significantly damages the ecosystem (INRENA 2008; Urrunaga et al. 2012, 2018). Illegal but highly profitable business opportunities in the frontier, like logging in Manu, tend to keep attracting an increasing number of people looking to profit from them and attempting to maintain them unchecked (Tsing 2009; Wilson 2004). Caused by the overwhelming illegality of the timber business in Peruvian Amazonia, contractual, labour and even sexual abuse are pointed out as common elements of this business (Leal et al. 2015; Salisbury et al. 2011; Urrunaga et al. 2012, 2018). The unfairness of trade and labour in this market is exacerbated by the ethnic hierarchies common in the frontier (Leal et al. 2015; Ráez Luna 2017; Tsing 2005; Urrunaga et al. 2012, 2018). Additionally, the high capital requirements of logging, its illegality, as well as its 'boom and bust' nature, preclude this business from generating significant local benefits (Barbier 2012; Harvey et al. 2017; Harvey and Knox 2008, 2015; Larsen 2016; Oliart and Biffi 2010; Sunderlin et al. 2005). Conversely, these illegal businesses tend to benefit more affluent people the most, giving the local elites with enough capital to delve into businesses like timber extraction a high economic and political power (Harvey and Knox 2015; Sunderlin et al. 2005).

The combined impact that both of these businesses have on social relations and local ecosystems can be significant for local livelihoods (Arima et al. 2005; Gallice et al. 2017; Sunderlin et al. 2005; Tsing 2009). For instance, in the absence of well-enforced property rights, the immigration incentivized by frontier roads has the potential to threaten the capital assets of preceding residents through increased possibilities of land encroachment and trafficking (AIDESEP 2018a; Araujo et al. 2011; Gould, 2006 cited in Bennett and Sierra 2014; Oliart and Biffi 2010; Sunderlin et al. 2005; Vergara et al. 2014). The increased access provided by roads, then, produces what Tsing calls a 'fugitive landscape', which drives Indigenous peoples and other preceding residents away from the resources they depend on for their livelihoods (2005, p. 39). Not surprisingly, these processes frequently motivate violent interactions and encounters, as well as vicious cycles of abuse within communities (Gallice et al. 2017; Oliart and Biffi 2010; Royo-Villanova 2017; Salisbury et al. 2011; Tsing 2005, 2009; Vadjunec et al. 2011). Given the gradual penetration of a frontier road, processes of this nature have long since been taking place in Manu (Gallice et al. 2017; INRENA 2008; Ráez Luna 2017). If completed all the way to Boca Manu, the road expansion currently proposed for this area is expected to generate uncontrolled immigration and land invasion (Gallice et al. 2017; Larrea-Gallegos et al. 2016; MINAM 2016).

Other businesses not dependent on intensive land use may seem to hold the actual benefits of economic connectivity but have pitfalls that must be considered as well (Brondo 2013). Like timber extraction, tourism is a capital-intensive business and, as such, remains a privileged source of income (Adams and Hutton 2007; Barbier 2012; Ráez Luna 2017). Moreover, despite its common image as a sustainable and innocuous alternative for areas like Manu, tourism may still have a considerable social and environmental impact, which could be intensified by road construction (Brondo 2013; Gallice et al. 2017; MINAM 2016). Additionally, the also vulnerable economic position that frontier dwellers are in often renders any attempts to incentivize them to shift away from established practices ineffective, particularly when focused on more 'sustainable' but less profitable alternatives (Adams and Hutton 2007; Larsen 2016).

When evaluating the social and economic consequences of the changes in livelihoods incentivized by road building in the Amazon, it is important to consider their environmental costs as well (Barrantes et al. 2014; Vergara et al. 2014). The livelihoods of Indigenous people, which often preserve certain 'traditional' components, depend the most on a well-preserved forest ecosystem (Sunderlin et al. 2005). On the other hand, migrants, attracted by the low barriers of entry to the forest frontier economy, usually base their livelihoods on the exploitation of forest resources through practices that are more damaging to ecosystems (Sunderlin et al. 2005). Because of the processes unleashed by colonization, the access provided by roads in Amazonia tends to promote dramatic land use change (Araujo et al. 2011; Carvalho et al. 2002; Pieck 2011; Vergara et al. 2014). The most immediate environmental effect of the expansion of road networks in Amazonia is forest fragmentation, followed very quickly by degradation and deforestation as consequences of logging and agriculture (Arima et al. 2005, 2013; Gallice et al. 2017; Urrunaga et al. 2012, 2018; Vergara et al. 2014). The cases of dramatic deforestation caused in the areas surrounding the communities connected by the Interoceanic Highway in Madre de Dios are a clear example of this (AIDESEP 2018a; Vergara et al. 2014). In the specific case of the unpaved road into Boca Manu, the few studies conducted predict an increase in carbon emissions, forest fragmentation and more than 43 thousand hectares of deforestation by 2040 (Gallice et al. 2017; Larrea-Gallegos et al. 2016; MINAM, 2016).

Despite the potential that properly enforced environmental regulations and property laws could have in curbing these effects, there seems to be a significant lack of political will to apply them effectively and a series of legal and political incentives that contradict them (Araujo et al. 2011; Carvalho et al. 2002; Global Witness 2019; INRENA 2008; Larsen 2016; Urrunaga et al. 2012, 2018). This reveals the superior priority that economic growth incentives often have over any environmental protection measures in policy and law enforcement (Gudynas 2016b). In this way, processes of internal migration and activities like logging generate deforestation and degrade the forest but have been part of the process of economic growth in the frontier, which has benefited local elites in frontier areas like Manu for decades (Gallice et al. 2017; Sunderlin et al. 2005; Tsing 2009). Thus, it seems that most economic benefits that roads are expected to promote are ultimately appropriated by extra-local agents or local elites, while most of its costs would be assumed by local populations of Indigenous peoples, previous landless migrants and the ecosystems they inhabit (Adams and Hutton 2007). Making matters worse, environmental and social costs tend to interact and form vicious cycles by which local communities remain poor and keep resorting to resource exploitation, which never produces significant benefits for their wellbeing (Figallo and Vergara 2014; Larsen 2016; Vergara et al. 2014).

As shown for each of these promises of road infrastructure, the unpredictability of the frontier prevents the outcomes aspired to from taking place. However, it is precisely the continued absence of the expected results what fuels the 'enchantments' of roads (Harvey and Knox 2012, 2015). Despite the potentially pernicious social impact and the environmental degradation that roads can generate, the 'enchantments' of infrastructural modernization motivate the prioritization of road construction over any potential cost (Gallice et al., 2017; Harvey and Knox 2012; Jiménez 2017; Salisbury et al. 2011). The unquestioned expectations of development that roads generate give way to the disregard of potential risks, which further disfavours frontier residents in comparison to incomers (Harvey and Knox 2008, 2012, 2015; Murray-Li 2014;

Uribe 2017). This is the case for both local populations, who can witness the failure of roads firsthand, and for decision-makers at various levels, who incorporate the dismissal of risks into the legal framework (Barrantes and Glave 2014; Dourojeanni et al. 2010; Gallice et al. 2017; Ráez Luna 2017; Vergara et al. 2014).

Given its normative characteristics, the influence of the discourse of development plays an important role in the entrenchment of aspirations of modernization in the minds of frontier dwellers and authorities alike (McEwan 2009; Taylor 2007). The notion of development through infrastructure is rooted in modernization dichotomies, by which the frontier is in stark contrast with civilized society and, as such, needs to improve (Taylor 2007; Uribe 2017). Moreover, the territorial implications of road building in the frontier, and its consequential colonization, contribute to the embedding of power relations that put pressure in favour of measures that will serve local elites (Bluwstein and Lund 2018; Peluso and Lund 2011). This is so even when viable alternatives that would be less risky for local communities and landscapes exist (Barrantes et al. 2014; Dourojeanni et al. 2010; Gallice et al. 2017; Vergara et al. 2014). As in other frontier contexts, risks seem to be acknowledged only when their dire consequences have taken place (Tsing 2005).

After considering the nature of the 'enchantments' of roads, or their capacity to maintain aspirations that support their construction regardless of their inability to fulfil them, we now turn to other aspects that influence the demands for road construction in Manu. The influence of power relations and territorial aspects in the entrenchment of the promises of infrastructure and the power of roads to 'enchant' are recognized but somewhat downplayed in the work of Harvey and Knox (2012). As will be seen in the following sections, these are deployed through territorialization, the use of the discourses of development and other political tactics, which in the case of Manu seem to have a more significant role than what Harvey and Knox recognize in their case study. As will be shown, the existence of protected areas adjacent to the road in question is of importance.

3.4 Territory and Power in Manu

Having evaluated how aspirations for modernization held by frontier dwellers foster the capacity of roads to 'enchant' and appear as unquestionable sources of progress, it is appropriate to take a deeper look at the political backdrop that underlies these issues (Bryant and Bailey 2000; Harvey and Knox 2012; Robbins 2004). Perhaps given their focus on the process of the Interoceanic Highway at a larger scale, Harvey and Knox seem not to pay much attention to the aspects described in the following sections (2012). We now turn to evaluate the territorial politics of rainforest conservation and regional infrastructural development, to analyse local power relations, as well as examine them through a postcolonial lens.

3.4.1 The Territorial Clash Between Conservation and Modernization

Given the existence of two protected areas in the Manu province, the MNP and the ACR, as well as several frontier communities, views of frontier modernization and conservation are in constant conflict with each other in this area (INRENA 2008; Oliart and Biffi 2010; Ráez Luna 2017; Shepard et al. 2010). Entities concerned with conservation, like SERNANP and various environmental NGOs, seem to view poverty as a constraint for conservation and tend to address matters of local well-being only as a means to attain a well-conserved ecosystem, which only produces short-term solutions (Adams et al. 2004; Ráez Luna 2017; Shepard et al. 2010). Regarding this, it must be considered that the isolation generated by the lack of road access has been an important pillar in the conservation of the MNP, the 'crown gem' of Peruvian conservation (Gallice et al. 2017; SERNANP 2013; Shepard et al. 2010). On the other hand, most elected and appointed authorities, political figures and communities involved seem to believe that conservation should not become an obstacle for economic growth through modernization, in which roads are believed to play an important role (Adams et al. 2004; Gallice et al. 2017; Harvey and Knox 2015; Ráez Luna 2017). Roads also play a role as modernizing infrastructures that allow subalternized[3] Indigenous Amazonian peoples to progress in the same path taken by now 'modern' Andean westernized society, of which most government officials in Manu are part of (Escobar 2008; Grosfoguel 2007; Harvey and Knox 2015; McEwan 2009; Mignolo 2000). These political forces, those which support environmental conservation and those that promote modernization through road building, have a significant influence on territory in Amazonia (Peluso and Lund 2011).

When considering the political impact of conservation measures in the Peruvian Amazon, the concept of the 'postfrontier' becomes relevant (Larsen 2016). According to Larsen (2016), this is a stage that follows the resource extraction frontier by adding 'sustainability measures', which appear to protect ecosystems but ultimately create further pressure on local resources and society instead of reducing them. The extraction of resources that essentialist frontier narratives give way to is perceived as a threat to the ecosystems that generate their abundance, which has motivated the application of countering environmental and social policies in the last few decades (Larsen 2016). However, given the imperative that underlies extraction in frontier spaces, most of these measures only respond symbolically to global pressures towards the adoption of 'green' policies and ultimately, even purposefully, have limited impact (Arsel et al. 2016a, b; Larsen 2016). Rather, the often incongruent measures that seek to curb the effects of uninhibited extraction in the frontier create further pressures on local resources and communities instead of reducing them, giving way to a new 'stage' of the frontier process (Larsen 2016). So, it is important to consider that the postfrontier stage does not mean the end of resource extraction in isolated rural areas but a mere re-structuring of its regulations, procedures and allowances (Larsen

[3] Here we use this term in the sense described by Mignolo, who emphasizes the performative nature of 'subalternization', i.e. being made a subaltern by colonial powers (2000).

2016; Tsing 2005, p. 32). The postfrontier, thus, introduces a new paradox to frontier spaces, one in which increasing environmental regulations and interventions contrast with the continuation of environmental degradation and the fragility of Indigenous rights (Larsen 2016). The hegemony of resource extraction over social and environmental concerns in the frontier is maintained, which is another manifestation of the entrenchment of economic growth as development (Gudynas 2016b; Larsen 2016; Salisbury et al. 2011). Nonetheless, the lack of effectivity of postfrontier measures does not reduce the animosity that frontier resource users and extractors have for them (Raycraft 2018). The postfrontier measures affecting Manu, the imposition of the Native Community framework and the establishment of two protected areas, are examined below considering their tokenistic nature and paradoxical impact (Oliart and Biffi 2010; Ráez Luna 2017; Shepard et al. 2010).

3.4.1.1 Indigenous Communities and Protected Areas as Tools of Territorialization

As mentioned in previous chapters, the 'Native Community' framework was established in 1974 and later modified in 1978 (Greene 2009; Larsen 2016; Shepard et al. 2010). This set of measures sought to legally recognize the territories that Indigenous Amazonian communities inhabited, a key factor to ensure access to the resources they based their livelihoods on and to preserve their self-determination (Chase Smith and Salazar 2014; Clech Lam 2004; Figallo and Vergara 2014). Notwithstanding, this framework was based on the 'Peasant Community' model, designed for Andean communities, and imposed the Andean agricultural economy over the traditional hunting, gathering and swidden agriculture lifestyle predominant in Amazonian cultures (Chase Smith and Salazar 2014; Greene 2009; Larsen 2016; Ohl et al. 2008; Oliart and Biffi 2010; Peluso and Alexiades 2005; Salisbury et al. 2011; Shepard et al. 2010). In this way, the semi-nomadic livelihoods of the Indigenous Amazonian peoples of Peru were pushed towards agriculture and, thus, to a sedentary lifestyle confined within bounded territories using the Andean agricultural model as a template (Greene 2006, 2009; Larsen 2016; Salisbury et al. 2011). Given the additional pressure of surrounding extraction, Indigenous Amazonian peoples were prompted by enclosure to legally claim whatever territories the state allowed them to, in order to salvage as much of their land as they could from outside influence (Larsen 2016; Peluso and Lund 2011; Salisbury et al. 2011).

The top-down inclusion into this legal framework limited the economic, administrative and political agency of Amazonian Indigenous communities upon their territories (Figallo and Vergara 2014). These communities were not given full ownership of the land but instead were granted its use by the state, which kept the ultimate decision over what these territories would be used for (Greene 2006, 2009; Leal et al. 2015). In this way, local communities were only granted certain 'benefits' from the land, while its actual property was left in the hands of the state and extractive elites (Oliart and Biffi 2010; Tsing 2005). This left Indigenous peoples on uneven ground for any territorial struggles they might face (Leal et al. 2015). The position that Indigenous Amazonian communities were placed on, thus, favoured the 'colonization' by Andean migrants that the state had been promoting, and further rooted the view of Amazonian peoples as socially inferior (Larsen 2016; Taylor 2007). Furthermore, considering the tokenistic nature of the Native Community framework and the ultimate decision power of the state over their territory, the facilitation of resource extraction seems to be the main reason behind the forced sedentarization of these communities (Leal et al. 2015; Peluso and Alexiades 2005). This postfrontier measure weakened Indigenous territoriality instead of bolstering it and pressured Indigenous groups throughout Peruvian Amazonia to adapt their traditional livelihoods to a new structure imposed from the outside (Oliart and Biffi 2010; Ráez Luna 2017). Consequentially, Indigenous Amazonian livelihoods were pushed away from their 'traditional' practices, which are precisely the ones conservation still counts on through a fetishized view (Adams and Hutton 2007; Kohler and Brondizio 2016; Oliart and Biffi 2010; Ráez Luna 2017; Tsing 2005). This is an important point of interaction between this framework and the creation of protected areas.

The establishment of the MNP, at around the same time as the initial creation of the Native Community framework, also reflects the social repercussions of the post-frontier (Larsen 2016; Ráez Luna 2017; Shepard et al. 2010). This protected area was created as if no human population lived within or around it (Larsen 2016; Ráez Luna 2017; Shepard et al. 2010). Thus, there were, and still are, serious discrepancies between Park regulations and the priorities and world vision of the communities influenced by it (Ráez Luna 2017; Shepard et al. 2010). Even legally, Peruvian protected areas only seek to preserve those Indigenous cultural aspects deemed as 'associated' with conservation, according to their authorities (Brondo 2013; Díaz and Miranda 2014; SERNANP 2014). UNESCO's declaration of the MNP and its buffer zone as a Biosphere Reserve in 1977 only increased the area where zoning and restrictions were applied (Adams and Hutton 2007; Shepard et al. 2010). This is a clear example of the capacity of protected areas to threaten and displace Indigenous peoples by enclosing them in territories where their livelihoods are restricted (Adams and Hutton 2007; Peluso and Lund 2011; Pieck and Moog 2009; Tsing 2005). Moreover, the exclusion of local communities from the management of these areas keeps them from being able to influence the adaptation of the corresponding regulations to their current livelihoods and aims (Bluwstein and Lund 2018; Kelly, 2011 cited in Peluso and Lund 2011; Ráez Luna 2017; Raycraft 2018). Despite attempts to revert this through alternative conservation models, most participatory measures adopted by 'communal reserves' are mostly symbolic and still leave little space for contestation or debate (Bluwstein and Lund 2018; Oliart and Biffi 2010; Raycraft 2018; Robbins 2004). The ACR is an example of this, given that the communities symbolically involved in its management must present management and development plans to be approved by outsiders before implementing them (Díaz and Miranda 2014; Larsen 2016; Oliart and Biffi 2010).

These restrictions mostly seek to limit the use of natural resources by Indigenous people and other local residents, turning activities that were key components of their livelihoods before the establishment of protected areas into illegal activities (Oliart and Biffi 2010; Peluso and Lund 2011). This is justified by Western imaginaries related to notions of the 'noble savage' and an Edenic nature, which legitimize these efforts to 'freeze' Indigenous people and their livelihoods into an idealized state of preceding harmony with the environment (Igoe 2004; Peluso and Lund 2011; West 2006). These restrictions establish the image of local resource users as despoilers of the environment, and of the state and global conservation interests as their 'distant landlords' (Adams and Hutton 2007; Peluso and Lund 2011, p. 669; Tsing 2005). As with the Native Community framework, Indigenous Amazonian people living within or around protected areas in Peru are reduced to the category of beneficiaries and informants, instead of being owners and decision-makers (Larsen 2016; Oliart and Biffi 2010). Under these conditions, protected areas around the world tend to be significantly costly to local residents, in both monetary and non-monetary terms, which keeps them expectant for compensations that they seldom receive (Adams and Hutton 2007; Kohler and Brondizio 2016; Peluso and Lund 2011; Raycraft 2018). Peruvian protected areas are not the exception, and often have a negative effect on household economic wellbeing, despite communal participation in their management (Barrantes and Glave 2014; Díaz and Miranda 2014; Ráez Luna 2017). The performance of Peruvian protected areas, which sometimes even increase deforestation due to superposition with extractive concessions, doesn't provide many arguments to justify these costs (Vergara et al. 2014). Mechanisms to revert these effects, through legitimate participation or economic compensation, are rarely in place (Adams and Hutton 2007; Ráez Luna 2017). Instead, the response to the failures of protected areas is to implement new conservation measures, assuming that their effects will be positive, which reveals the political power of conservation as opposed to the communities it tends to affect (Adams and Hutton 2007; Bluwstein and Lund 2018; Peluso and Lund 2011; Tsing 2000). In this way, and despite the undoubted importance of the conservation of ecosystems, the implementation of conservation measures frequently becomes entangled with the disenfranchisement and dispossession of local communities (Adams and Hutton 2007; Raycraft 2018; Robbins 2004).

The forceful modification of the interactions of Indigenous people with their territories can be considered an act of territorialization for conservation (Bluwstein and Lund 2018; Delaney 2005). The imposition of territorial management systems that prioritize Western views in favour of the state's goals and those of the global conservation movement is taken forward regardless of its effects on local livelihoods (Larsen 2016; Ráez Luna 2017; Raycraft 2018). But this is, in fact, the very nature of the original idea behind national parks, described eloquently in Tsing's analysis of John Muir's conservation philosophy (2005, p. 140): to privilege the 'cosmopolitan traveller', a potential park visitor (Adams and Hutton 2007). These asymmetrical relations have prompted the animosity of Amazonian Indigenous peoples and their organizations, as well as of communities established in buffer zones, towards international conservation since the inception of protected areas (Díaz and Miranda 2014; Larsen 2016; Pieck and Moog 2009; Ráez Luna 2017). This effect has also extended towards environmental law enforcement agencies (Global Witness 2019). This creates a state of tension that constantly generates pressure over the very ecosystems protected areas seek to conserve and prevents local communities from assuming responsibility for these ecosystems, which could constitute a kind of 'rebound effect' of conservation measures (Díaz and Miranda 2014; Larsen 2016; Oliart and Biffi 2010; Ráez Luna 2017; Robbins 2004; Tsing 2005). The aspiration that the communities of Manu have for road connectivity is a manifestation of this effect, which reinforces the capacity of the Manu Road to 'enchant' (Gallice et al. 2017; Harvey and Knox 2012). In the long term, this counterpressure leaves the management of Peruvian protected areas with few options apart from directly satisfying the necessities of these communities, which they are not legally responsible or moneyed for, or making significant concessions regarding their regulations (Gallice et al. 2017; Kohler and Brondizio 2016; Larsen 2016; Ráez Luna 2017; Shepard et al. 2010). The subsequent approval of the construction of the Manu Road by SERNANP can be considered one such concession (SERNANP 2017). This could be reverted in the long term by increasing legitimate and independent local participation, which has had positive effects in some Peruvian protected areas (Brosius 1999; Kohler and Brondizio 2016; Ráez Luna 2017; Royo-Villanova 2017). However, this would require breaking through the colonial inheritance of the system of protected areas and the tendency to respond to the failures of conservation measures by increasing them (Adams and Hutton 2007; Bluwstein and Lund 2018; Grosfoguel 2007; Peluso and Lund 2011; Robbins 2004).

As we have explained in this section, the postfrontier measures applied in Manu have ultimately worked against their stated purposes and acted as tools of territorialization (Bluwstein and Lund 2018; Delaney 2005; Larsen 2016). The Native Community framework has surreptitiously weakened Indigenous territoriality, facilitating extraction, while the establishment of protected areas has exacerbated pressure upon natural resources by restricting Indigenous resource use and influence in decision-making (Díaz and Miranda 2014; Larsen 2016; Leal et al. 2015; Oliart and Biffi 2010; Peluso and Alexiades 2005; Ráez Luna 2017; Robbins 2004; Tsing 2005). Apart from these territorializing forces, further contention for land control is driven by the expansion of the road network at a South American, continental, scale, which we will analyse in the following section.

3.4.1.2 Infrastructural Modernization as 'Neoliberal Territorial Design'

It must be kept in mind that the postfrontier stage does not signify the end of forest degradation or modernist development but their continuation under ultimately symbolic environmental protection measures (Larsen 2016; Tsing 2005). Thus, the implementation of postfrontier measures in the Amazon has not stopped the expansion of large infrastructure through projects like IIRSA (González 2012; Kanai 2016; Pieck 2011, 2013; van Dijck 2008; Zhouri 2010; Zibechi 2006). This ongoing project seeks to integrate the continent through the construction of large infrastructure corridors that would help overcome its many 'natural barriers' and allow international commerce to boost regional competitiveness and economic growth (González 2012; Jarrier 2012; Kanai 2016). As a response to the South American economic boom of the early 2000s, most states in the continent adhered to it, despite the admitted likelihood of it generating social conflict and environmental degradation through the drastic modification of connectivity and of land use incentives (Jarrier 2012; Pieck 2013; Zhouri 2010). Being significantly influenced by the modernist development policies that pervaded the continent in the twentieth century, its main focus was to redesign and extend territorial links in the region to cater to global market demands for commodities, without considering internal needs (Barrantes et al. 2014; Kanai 2016; Pieck 2013, 2015; van Dijck 2008; Zhouri 2010; Zibechi 2006). Thus, several authors point at IIRSA as a strategy that puts rich South American territories at the outright service of global markets (Pieck 2015; Quinhoes 2011; van Dijck 2008; Zhouri 2010; Zibechi 2006), what Kanai calls a 'pervasive form of neoliberal territorial design' (2016, p. 160). Despite this, the successive governments of countries like Peru have continued supporting IIRSA and its effects on key territories like Amazonia, increasing pressures over them and their inhabitants (Barrantes and Glave 2014; Oliart and Biffi 2010; Pieck 2013, 2015; Ráez Luna 2017).

The most significant manifestation of this regional strategy in Peru is the IIRSA Sur, better known as the Interoceanic Highway, which connects the southern Peruvian coast with Brazil through the southern Peruvian Andes and Amazonia (Barrantes et al. 2014). It has been a symbol of modernization for the south of the country, especially for the Madre de Dios region, as it is the first major road that goes through it (Barrantes et al. 2014; Harvey and Knox 2015; Pieck 2013, 2015). It has also been the locus of unprecedented investments in transport infrastructure, justified by increased connectivity with the Brazilian economy and access to the ports of the Pacific for Brazil (Barrantes et al. 2014; Pieck 2013). However, its construction has not been able to fulfil its integratory objectives or to avoid the significant social and environmental consequences of providing increased access to natural resources and expanding the frontier (Harvey and Knox 2015; Kanai 2016; Pieck 2013, 2015). The major causes of these failures have been the lack of consultation procedures and the explicit avoidance of environmental evaluations through a process of legal prioritization implemented by the state (Harvey and Knox 2015; Pieck 2013, 2015). Zibechi calls this avoidance of consultation and of other checks and balances the 'silent' construction of IIRSA (2006). The 'silent' construction of large components of IIRSA prompts the construction of smaller infrastructures that seek to join its larger network in a similarly unchecked manner. This also facilitates the avoidance of accountability for the effects of the network as a whole (Zibechi 2006). This effect, thus, could be facilitating the expansion of IIRSA's 'neoliberal territorial design' into previously unanticipated territories through 'silent' side-projects for smaller infrastructures (Kanai 2016; Zibechi 2006).

The most recent extension of the Manu Road can be considered one such case: a smaller road that seeks to join the Interoceanic Highway and that has been built without proper consultation, avoiding environmental assessments and incurring in illegal proceedings (AIDESEP 2018a; Gallice et al. 2017; Larrea-Gallegos et al. 2016; Ráez Luna 2017). What Gallice et al. call the 'incremental expansion' of the Manu Road, the building of smaller stretches as individual projects instead of a one-off full extension, has facilitated the concealment of its overall effects, thus facilitating its legal approval (2017, p. 6). Furthermore, the attempt to declare this road as a 'public necessity' through the National Congress sought to override legal checks and balances established for this type of project (AIDESEP 2018a; Congreso de la República 2013; Ráez Luna 2017). Even if this declaration was denied by the National Government, the declaration of the road as such by GOREMAD gave it enough legitimacy to take it forward (Gallice et al. 2017; Humala and Cateriano 2015; Larrea-Gallegos et al. 2016; MINAM 2016). In fact, several of the other 'silent' infrastructure projects mentioned at the beginning of this chapter have been similarly proposed in the wake of the construction of the Interoceanic Highway (AIDESEP 2018b; El Peruano 2017; SPDA 2017a; Zibechi 2006). The replication of this mechanism to promote other similar projects has prompted the Inter-ethnic Development Association of Peruvian Amazonia[4] (AIDESEP), to declare such a mechanism as a 'perversion'[5] of this legal recourse (AIDESEP 2018a). Other inter-ethnic organizations also protest the lack of proper consultation proceedings in the expansion of the Manu Road, which is another coincidence with the construction of the Interoceanic Highway (Gallice et al. 2017; Harvey and Knox 2015; Pieck 2013, 2015). Moreover, the fact that the Manu Road is likely to be of most use to the extraction of timber and other resources, albeit illegally, links it to IIRSA's purpose of facilitating access to commodities to international markets and the neoliberal territorial design that this regional project follows (Gallice et al. 2017; Kanai 2016; Larrea-Gallegos et al. 2016; MINAM 2016; Zibechi 2006). Adding to this effect is the Peruvian National Government's lack of capacity to stop GOREMAD from going forward with the construction of the road (Ráez Luna 2017). Considering SERNANP's later approval of this road as 'compatible' with MNP and ACR objectives, this lack of capacity could also be understood as a lack of willingness (SERNANP 2017). This would follow what Larsen calls 'the paperwork reserve syndrome',[6] the continued prioritization of extraction over the enforcement of protected area regulations (Larsen 2016, p. 129). The use of these means to expedite the construction of the Manu Road, and this road's apparent capacity to bring the economic growth promised by IIRSA closer to the communities it connects, further reinforces the appearance of incoming prosperity (Harvey and Knox 2012; Kanai 2016; Tsing 2000, 2005; Zibechi 2006).

This section has established the influence of IIRSA upon the expansion of the Manu Road and drawn key parallels between both projects despite differences in scale, which indicate the Manu Road is a continuation of IIRSA's territorial aims. As the following section emphasizes, the confrontation between these and the territorial effects of conservation ultimately favour the expansion of this infrastructure.

---

[4]*Asociación Interétnica de Desarrollo de la Selva Peruana.*

[5]'*Desnaturalización*' (AIDESEP 2018a, p. 19).

[6]'*Síndrome de la reserva en papel*' (Larsen 2016, p. 129).

3.4.1.3 Confronted Territorializations

Both the conservation sector, through the establishment of restrictive protected areas where local populations have little input, and extractive interests, through the construction of roads that overlook legal requirements to serve the global demand for commodities, territorialize over the same area (Bluwstein and Lund 2018; Delaney 2005; Kanai 2016). This becomes an example of how powerful actors enter in conflict with each other in their struggle for land control, which can have significant effects on the livelihoods of less powerful local populations (Bryant and Bailey 2000; Peluso and Lund 2011). However, even when appearing as diametrically opposed, MNP and ACR's conservation for territorialization and IIRSA's neoliberal territorial design share a territorial logic (Baletti 2012). As Baletti (2012) explains, environmentalism and developmentalism are both focused on assigning property rights to achieve the optimal results of effective conservation or maximum economic growth, respectively. These property rights, reflected in the regulations, incentives and concessions that shape sustainable and modernizing development on the ground, are assigned to specific actors, deemed as 'deserving' of these (Baletti 2012). This reflects a strong will to impose these types of development from the outside given that these come from what are regarded as superior types of knowledge than those that could be produced locally (Cupples et al. 2019; McEwan 2009). The common roots in Western thinking of both conservation and modernization are revealed through the North to South administration of territory (Adams and Hutton 2007; Cupples et al. 2019; Grosfoguel 2007; McLeod 2007).

So, both of these apparently confronting forces attempt to turn a disputed territorialized space into a global good that creates value for distant outsiders (Peluso and Lund 2011). The case of Peruvian modernizing thought is epitomized in this sense by ex-president[7] Alan García's 'Dog in the Manger'[8] philosophy (Chase Smith and Salazar 2014; Jiménez 2017; Larsen 2016; Royo-Villanova 2017). Even though this philosophy is based on decades-old concepts of modernization, including president Belaúnde's justification for the 'colonization' of the Peruvian Amazon, it has guided Peruvian development strategies since the late 2000s (Ames 2010; Jiménez 2017; Larsen 2016; Morel 2014). Its rationale holds that Indigenous groups are undeserving occupants of resource-rich territories that remain idle under their management when these could be generating economic growth that would serve the whole country through the exploitation of resources and commodity exports (Chase Smith and Salazar 2014; Jiménez 2017; Larsen 2016; Royo-Villanova 2017). Thus, Indigenous peoples are mandated to give up their territories, even forced through violent means, in order to facilitate the country's economic growth and to satisfy global commodity

---

[7]His second mandate lasted from 2006 to 2011.

[8]'El *Perro del Hortelano'* is a reference to a popular Western fable of a farmer's guard dog who doesn't allow other animals to eat from the manger, even when the dog was not going to eat from it himself. García's use of this fable as a metaphor was not perceived well by Indigenous Amazonians, who felt they were being called dogs outright (Royo-Villanova 2017).

needs, which have seldom benefited them directly (Larsen 2016; Morel 2014; Royo-Villanova 2017). On the other hand, 'sustainable development', on which current conservation policies are based on, regards tropical forests as 'global public goods', as they provide climate stabilization services for the whole planet (Barrantes and Glave 2014). The specific purpose of protected areas under this logic is also to preserve resources for future generations of Peruvians, while providing mere 'benefits' to local populations (Díaz and Miranda 2014; Larsen 2016; Oliart and Biffi 2010; Tsing 2005). The stake of the Peruvian government is the upholding of international climate change agreements and ensuring the general population is benefitted by a well-conserved environment (Gallice et al. 2017; Humala and Cateriano 2015; MINAM 2016). This imposes the vision of 'what should be' held by global conservation, mostly represented by privileged foreigners, upon subalternized ethnic groups, questioning them if they do not adjust to it (Adams and Hutton 2007; Larsen 2016; Mignolo 2000; Peluso and Lund 2011; Raycraft 2018; Tsing 2005; Vadjunec et al. 2011).

In this way, modernization and conservation policies, enacted by different state instances, conform a double-sided source of pressure upon Indigenous Amazonian peoples and their territories (Pieck and Moog 2009). Considering territorialization is a disempowering force for local communities, the fact that its sources are twofold in this case makes its effects even more serious (Delaney 2005). Amazonian Indigenous communities often seek to adapt to these multiple sources of pressure, which can reach overwhelming extremes (Leal et al. 2015; Ráez Luna 2017). The results of this confrontation of territorializing forces may finally favour the capacity of roads to 'enchant'. As the pressures of conservation forebode limitations to economic prosperity through further regulation in favour of extra-local interests, the communities affected by them tend to reject them (Díaz and Miranda 2014; Larsen 2016; Oliart and Biffi 2010; Ráez Luna 2017; Robbins 2004; Tsing 2005). To the contrary, the entrenchment of the principles of economic growth and modernization, as well as the capacity of capitalism to proliferate in the frontier and control territories that hold valuable resources, have the potential to tilt the balance towards these (Escobar 2008; Gudynas 2016a; Tsing 2005). The weakness of Indigenous territorial resources does not make resistance to these forces easy and Indigenous leaders are often tempted to 'join them' when feeling they cannot 'beat them', becoming 'enchanted' by modernizing infrastructures (Larsen 2016).

This section has analysed the various aspects of territoriality in Manu and explained how their combined effects ultimately motivate local support for the expansion of the Manu Road. The alignment of the apparent benefits of IIRSA with the aspirations for modernization held by the communities in Manu stands in contrast to the restrictions imposed by conservation, which seem to stagnate livelihoods for the benefit of outsiders. In the following two sections, we explore how power relations in Manu and the colonial implications of several elements of this case further reinforce local support for this infrastructure.

3.4.2 Power Relations in Manu

The push for modernization can manifest itself in the frontier through the influence of powerful actors like local politicians, who take credit for road construction, and local extractive elites, who are likely to benefit the most from roads (Harvey et al. 2017; Harvey and Knox 2015). Their social and political influence adds to the overwhelming aspirations that frontier dwellers have, while their power to represent non-elites as 'others in crisis' facilitates the justification of road construction (Campbell 2012; Harvey and Knox 2012; McEwan 2009). Through this influence, and despite their potential to fail to increase wellbeing and their likely risks, roads continue to be considered an ultimate means for improving wellbeing in the frontier (Campbell 2012; Carvalho et al. 2002; Dalakoglou and Harvey 2012; Harvey and Knox 2012, 2015; Sunderlin et al. 2005). This further motivates communities to adamantly demand road connectivity and even to force those uninterested or opposed to it to do so as well (Harvey and Knox 2012, 2015). Thus, local power relations and the capacity of actors to influence decision-making become an important element in the analysis of the 'enchantments' of roads (Bryant and Bailey 2000; Harvey and Knox 2012).

In general, infrastructures tend to favour more powerful and wealthier actors, and so do social structures near protected areas (Adams and Hutton 2007; Harvey et al. 2017). The unequal distribution of power allows elites to appropriate benefits in these and other contexts (Bryant and Bailey 2000). The power that elected and appointed authorities have to enforce the law and invest public funds in accordance with their political and economic motivations turns them into territorializing actors (Lemos and Agrawal, 2006 cited in Brondo 2013; Peluso and Lund 2011). In the Peruvian case, recent decentralization policies have given more power to such authorities and mixed with prevalent corruption, which has revealed the economic motivations that authorities in Manu and other Amazonian provinces have (Barrantes and Glave 2014; Gallice et al. 2017; Salisbury et al. 2011). Some cases have also revealed the capacity of economic elites, namely illegal loggers, to appropriate the benefits of road construction (Gallice et al. 2017; Larsen 2016; MINAM 2016). Loggers gain the most from these infrastructures given the increased access to hardwoods that these allow and their low surface quality, which few except them are able to use profitably (Gallice et al. 2017; Larsen 2016). The numerous presence and economic power of loggers in the frontier is just another asset to their political influence (Gallice et al. 2017; Harvey and Knox 2015; INRENA 2008).

Several aspects of governance in the Peruvian Amazon facilitate this (Larsen 2016). According to Larsen's analysis of the postfrontier in the central Peruvian Amazon, this appropriation of benefits by local elites is facilitated by what he calls 'inverse governance'[9] (2016, p. 149). This term describes the way in which legal frameworks established to reduce deforestation and facilitate Indigenous territorial control are easily manipulated by extractors to facilitate exactly the opposite purposes (Larsen 2016). Guided surreptitiously by the 'dog in the manger' principles described previously, regulations of this sort, which have important territorial implications,

[9] '*Gobernanza inversa*' (Larsen 2016, p. 149).

are purposefully established as tokens that do not fulfil their manifest objectives (AIDESEP 2018a; Jiménez 2017; Larsen 2016; Vergara et al. 2014). Additionally, the inverse effects of these perfunctory measures are exacerbated by ineffective law enforcement, which is often also deliberate, especially when actuated by regional governments (Barrantes and Glave 2014; Global Witness 2019; Jiménez 2017; Oliart and Biffi 2010). The subalternization of Indigenous Amazonians as inferior 'others' and the colonial aspects of the legal frameworks that their communities and territories are governed by underlies these effects (Jiménez 2017; McEwan 2009; Mignolo 2000; Ráez Luna 2017). In the case of Manu, this is evidenced by the apparent capture of power in certain communities through the abuse of the Native Community framework that several organizations warn about (Gallice et al. 2017; Oliart and Biffi 2010). Furthermore, the reduced access to cash that prevails in Native Communities comes together with the potential for political and economic rewards that their chiefs are offered, which is likely to foster support for powerful actors and their interests (Larsen 2016; Leal et al. 2015; Oliart and Biffi 2010; Ráez Luna 2017; Urrunaga et al. 2012, 2018). Thus, the conditions are set for powerful economic and political actors to use their influence and ineffective legal frameworks to reap the highest benefits of modernizing infrastructures like first roads.

Apart from said conditions, the likelihood of power capture and the overwhelming support for road construction from communities in Manu despite its evident effects speak of the capacity of powerful actors to 'conjure' the apparently ideal results of road building (Gallice et al. 2017; Oliart and Biffi 2010; Ráez Luna 2017; Torres López 2016; Tsing 2000, 2005). According to Tsing, acts of 'conjuring' are used to portray modernization and other cosmopolitan ideals as indisputable generators of wellbeing in order to secure external funding and local support (2000, 2005). The power of conjuring allows the portrayal of cosmopolitan universals as homogenous in specific contexts, thus perfectly attainable locally, which gives the promises of infrastructure an increased capacity to become 'enchantments' (Harvey and Knox 2012, 2015; Tsing 2000, 2005). Through the reproduction of discourses that universalize the binaries of modernization, as well as views of frontier dwellers as an 'other' in crisis, an 'economy of appearances' is created (McEwan 2009; Tsing 2000; Uribe 2017). In order to do this, according to Tsing, a yet nonexistent economic prosperity is 'conjured' in order to attract financial investment (2000, 2005). In the case of the Manu Road and other infrastructures, it seems that this financial investment is replaced by popular buy-in for road construction among local communities (Gallice et al. 2017; Harvey and Knox 2012; Oliart and Biffi 2010; Ráez Luna 2017). There, the power of conjuring seems to be strong enough to generate violent resistance to any attempts to oppose this infrastructure (Gallice et al. 2017; Torres López 2016). However, considering the work of Oliart and Biffi in the ACR (2010), as well as that of Larsen in the central Peruvian Amazon (2016), it is likely that the communities in Manu are not passive victims of 'conjuring' in favour of the road and the resource extraction it will facilitate. Their work shows several examples in which Indigenous Amazonian people contest the power of the extractive elites that threaten their territories by seeking to conduct the same extractive activities themselves, thus reclaiming the profits that their own territorial resources can provide. Similar examples of Indigenous agency are likely to still be taking place in the Indigenous communities related to the ACR and show that the power of 'conjuring' by logging elites that promote the Manu Road is not absolute (Oliart and Biffi 2010; Tsao 2016; Tsing 2005; Walker 2012).

In this section, we have explored the nature of the power relations present in Manu, the power of local elites to 'conjure' the benefits of road expansion and to force support for it, aided by the fragile governance structures present in Manu and the rest of the Peruvian Amazon. In all, even when their power to influence the road expansion process is not absolute, the end result is highly favourable for the modernizing project in Manu. Additionally, the existence of territorializing forces and the presence of powerful incomers in Manu is significantly influenced by colonial power structures that remain in the Amazonian frontier in Peru, which we analyse in the final section of this chapter.

3.4.3 A Postcolonial Analysis

The fact that most of the processes surrounding the case of the Manu Road are influenced by the will of extra-local forces and the arrival of people seen as ethnically superior to Indigenous Amazonians, a look at the issue through a postcolonial lens is useful to clarify their implications (Ames 2010; Cupples et al. 2019; Figallo and Vergara 2014; McEwan 2009; McLeod 2007; Mignolo 2000). The vision of the Amazon as '*Terra Nullius*', an uncivilized territory empty of the apparent advantages of Western society and therefore in need of their presence, is crucial in this analysis (McEwan 2009). This notion is the basis for the myth of the 'empty Amazon', through which Amazonian territories are seen as unpopulated and freely available land (Barbier 2012; Figallo and Vergara 2014; Hecht, 2004 cited in Salisbury et al. 2011). This also entails a vision of Amazonian peoples as 'savages' that need to be civilized and, consequentially, of frontier territories as politically empty and in need of a political structure that would 'improve' social interactions within them (Ames 2010; Figallo and Vergara 2014; McEwan 2009; Tsing 2005). When viewed as such, Amazonian peoples are also vulnerable to the subalternization of their knowledge by more powerful outsiders who see themselves as superior benefactors (McEwan 2009; Mignolo 2000). Furthermore, frontier territories seen as 'empty' of civilized people are also conceived as 'wild' and unproductive pools of resources waiting to be made profitable by being placed at the disposal of modern economic systems (Harvey and Knox 2015; Murray-Li 2014). These aspects are evident in the power structures found in Manu, and in the processes of colonization and territorialization that have taken place there, as well as in the arguments that justify them. Nonetheless, the history of colonization in the Peruvian Amazon and the power of current modernizing forces in Manu are likely to have motivated the emergence of opposing voices in response to them (Cupples et al. 2019; Escobar 2008). As the testimonies I gathered show, these voices are especially likely to emerge in on-road communities where the road has already had an effect, which will be explored and discussed in following chapters.

The process of purposeful 'conquest' of the Amazon in the 1950s and 60s carried with it a significant colonial inheritance that goes beyond considering the Amazon rainforest as empty, '*Terra Nullius*' or a pristine Eden (Adams and Hutton 2007; McEwan, 2009; Morel 2014). The fact that Andean people, deemed as subaltern but nonetheless legitimate citizens of Peru, were encouraged to settle in the territories of 'savage' Indigenous Amazonian peoples to facilitate state control over resources can be considered a process of internal colonization (Ames 2010; Barrantes et al. 2014; Grosfoguel 2007; Larsen 2016; McEwan 2009; Morel 2014; Pieck 2011). Furthermore, the state's purpose of putting these resources at the service of national urban elites and the foreign demand for commodities is a continuation of colonial structures that designated Indigenous territories as mere sites of extraction (Jiménez 2017; McEwan 2009; Vadjunec et al. 2011). Finally, the internal colonization of Indigenous Amazonian peoples, mandated by the state, and the consequent embeddedness of their subalternization is another relevant legacy of this process which has had long-term consequences (Ames 2010; Cupples et al. 2019; Mignolo 2000; Morel 2014).

Furthermore, the colonial elements perpetuated on Peru's social and geographic structures help justify the subalternization of Indigenous peoples and their knowledge, which is the source of several matters relevant to this case (Ames 2010; Grosfoguel 2007; McLeod 2007; Mignolo 2000). The perceived inferiority of Indigenous knowledge is a key component of the desires for cosmopolitan consumption that supports the aspirations held by frontier dwellers, who feel they need to overcome their condition as inferiors through modernization (Harvey and Knox 2012; Mignolo 2000; Tsing 2005). Considering other types of knowledge as superior is also what prompts the process of acculturation of Indigenous Amazonians, who seek to adopt the lifestyles of the West (Figallo and Vergara 2014; Leal et al. 2015; Oliart and Biffi 2010; Ráez Luna 2017). The subalternization of knowledge also naturalizes the binaries of modernization and universal notions on which the promises of infrastructure are based on, making them unquestionable (Harvey and Knox 2012, 2015; McEwan 2009; Mignolo 2000; Tsing 2005; Uribe 2017). This plays a role in the lack of fulfilment of the promises of infrastructure, and in their preservation as 'enchantments' as well (Harvey and Knox 2012). The subalternization of frontier dwellers as inferior 'others' legitimizes their displacement through the violent effects of the conflicts prompted by increased road access and the immigration of Andean peoples in search for land and economic opportunities (Barbier 2012; Barrantes et al. 2014; Gallice et al. 2017; Leal et al. 2015; McEwan 2009; Pieck 2013; Tsing 2005, 2009; Uribe 2017; Vadjunec et al. 2011). This also disfavours frontier dwellers in economic dealings with incoming urban traders who have a higher negotiating power given their higher social standing (Leal et al. 2015; Murray-Li 2014; Ráez Luna 2017; Sunderlin et al. 2005; Tsing 2005; Wilson 2004). Ultimately, these frontier infrastructures allow economic elites, who are mostly of outside origin, to appropriate local resources and most of the benefits they are supposed to bring to local populations (Bryant and Bailey 2000; Harvey and Knox 2015; Murray-Li 2014). In general, incorporation into the larger Peruvian society as subaltern members puts frontier dwellers in a vulnerable position, instead of favouring their wellbeing (Harvey and Knox 2015; Murray-Li 2014; Pieck 2013, 2015; Tsing 2005).

The subalternization of knowledge and the concept of an 'empty Amazon' also legitimize the processes of territorialization that affect Manu and the territories of Indigenous Amazonian peoples in Peru (Barbier 2012; Figallo and Vergara 2014; Mignolo 2000; Peluso and Lund 2011; Salisbury et al. 2011). The establishment of the MNP as if it was unpopulated is a clear reflection of this, as the communities affected by it were initially invisibilized and more recently deemed unworthy of participating legitimately in the area's management, which ignores their agency (Bluwstein and Lund 2018; Larsen 2016; Peluso and Lund 2011; Ráez Luna 2017; Raycraft 2018; Shepard et al. 2010). The basis of this exclusion is the view of local resource users as depleters, which legitimizes the forbiddance of certain components of their livelihoods and generates considerable costs for their households (Adams and Hutton 2007; Barrantes and Glave 2014; Díaz and Miranda 2014). The imposition of a vision of conservation held by privileged Westerners upon subalternized groups in a non-Western nation, as well as the penalization of those who do not respect it, further reflects a pattern of domination (Adams and Hutton 2007; Larsen 2016; Mignolo 2000; Peluso and Lund 2011; Raycraft 2018; Tsing 2005; Vadjunec et al. 2011). On the other hand, the historical role of roads as modernizing infrastructures that allow subalternized Indigenous Amazonian peoples to 'improve′ by becoming more alike to the rest of the country further reflects a prevalent colonial legacy (Escobar 2008; Grosfoguel 2007; Harvey and Knox 2015; McEwan 2009; Mignolo 2000). Adding to this is the imposition of an agricultural economy upon previously semi-nomadic communities in order to push for their sedentarization and increase access to the resources in their territories (Ames 2010; Greene 2009; Larsen 2016; Oliart and Biffi 2010; Salisbury et al. 2011; Shepard et al. 2010). Moreover, the fact that the regional infrastructure network serves the interests of external and more powerful actors reveals the influence of colonial principles on the development strategies that affect Indigenous territories (Gudynas 2016a; Kanai 2016; McEwan 2009; Tsing 2005; Uribe 2017). The exclusion of Indigenous people from decision-making related to these infrastructures is another element of this colonial influence (Harvey and Knox 2015; Pieck 2013, 2015). Much of this neoliberal territorialization is legitimized by the 'Dog in the Manger' discourse, which has profound neo-colonial implications (Ames 2010; Jiménez 2017; Larsen 2016; McEwan 2009; Morel 2014). Its consideration of Indigenous peoples as unworthy of the territories they occupy because they do not put the resources in them to the service of Peru's economic growth reveals their categorization as second-class citizens (Chase Smith and Salazar 2014; Jiménez 2017; Larsen 2016; Royo-Villanova 2017). So, through the subalternization of local knowledge, the forces of conservation and capitalism attempt to turn a disputed territorialized space into a territory that creates value for distant outsiders (Mignolo 2000; Peluso and Lund 2011). In general, the North to South administration of territory puts the Western influence of both territorializing forces in evidence (Adams and Hutton 2007; Cupples et al. 2019; Grosfoguel 2007;

McLeod 2007). This, again, is a reflection of colonial structures by which the decisions of the state and Western conservation interests are intellectually and morally superior than those of Indigenous communities (Adams and Hutton 2007; Grosfoguel 2007; Jiménez 2017; McEwan 2009).

Overall then, the colonial structures still present in Peru have a significant influence on the processes of territorialization and the power relations that determine the origins and outcomes of the conflict generated by the eventual construction of the Manu Road. The relatively low power of Indigenous Amazonian peoples regarding local elites and extra-local political forces considerably reduces the influence of the former upon the road expansion process, ensuring larger benefits for the latter. The dominance of powerful actors, nonetheless, does not translate into the nonexistence of contesting voices (Cupples et al. 2019; Escobar 2008).

In this chapter, we have discussed the concept of 'enchantments of infrastructure', its relation to the process of 'colonization' of the Peruvian Amazon, the influence of territorialization in Manu on them, as well as that of local power relations and the perpetuation of colonial structures. Analysing these different angles of the conflict over the Manu Road allows for a more nuanced understanding of the processes, scales and actors at play here. Together, they form the fabric in which the voices and insights presented in the following chapters are woven. But first, in the next chapter, we will revise methodological approaches and considerations that shaped this research.

References

Adams WM, Aveling R, Brockington D, Dickson B, Elliott J, Hutton J et al. (2004). Biodiversity conservation and the eradication of poverty. Science 306(November):1146–1150

Adams WM, Hutton J (2007) People, parks and poverty: political ecology and biodiversity conservation. Conserv Soc 5(2):147–183

AIDESEP (2018a) Informe sobre la situación de los pueblos en aislamiento y contacto inicial de la Amazonía peruana. AIDESEP, Lima

AIDESEP. (2018b). Pronunciamiento de AIDESEP sobre la LEY Nº 30723. AIDESEP Website (30723):1–2. http://www.aidesep.org.pe/pronunciamiento-de-aidesep-sobre-la-ley-no-30723/

Ames P (2010) Desigualdad y Territorio en el Perú: una geografía jerarquizada. Argumentos, 1

Araujo C, Bonjean CA, Combes J, Araujo C, Bonjean CA, Combes J, Motel PC (2011) Does Land Tenure Insecurity Drive Deforestation in the Brazilian Amazon? CERDI, Etudes et Documents 2010(13):2–34

Arima EY, Walker RT, Perz SG, Caldas M, Arima EY, Walker RT et al (2005) Loggers and forest fragmentation: behavioral models of road building in the Amazon Basin. Ann Assoc Am Geogr 95(3):525–541

Arima EY, Walker RT, Souza C, Pereira R, do Canto O (2013) Spontaneous colonization and forest fragmentation in the Central Amazon Basin. Ann Assoc Am Geogr 103(6):1485–1501

Arsel M, Hogenboom B, Pellegrini L (2016a) The extractive imperative and the boom in environmental conflicts at the end of the progressive cycle in Latin America. Extr Ind Soc 3(4):877–879

Arsel M, Hogenboom B, Pellegrini L (2016b) The extractive imperative in Latin America. Extr Ind Soc 3(4):880–887

Baletti B (2012) Ordenamento territorial: neo-developmentalism and the struggle for territory in the lower Brazilian Amazon. J Peasant Stud ISSN 39(2):573–598

Barbier EB (2012) Scarcity, frontiers and development. Geogr J 178(2):110–122

Barrantes R, Fiestas J, Hopkins Á (2014) Evolución de la infraestructura de Transporte y Energía en la Amazonía Peruana. In: Barrantes R, Glave M (eds) Amazonía Peruana y Desarrollo Económico. GRADE, IEP, Lima, pp 109–160

Barrantes R, Glave M (2014) Introducción. In: Barrantes R, Glave M (eds) Amazonía Peruana y Desarrollo Económico. GRADE, IEP, Lima, pp 13–20

Bennett DE, Sierra R (2014) Multi-scale dimensions of Indigenous land tenure in the Amazon. Hum Ecol 42:551–563

Bluwstein J, Lund JF (2018) Territoriality by conservation in the Selous-Niassa Corridor in Tanzania. World Dev 101:453–465

Bonifaz, R. (2018). Organizaciones Indígenas y aliados de Brasil se pronuncian sobre la construcción de carreteras en la Amazonía. Retrieved from http://www.aidesep.org.pe/organizaciones-indigenas-y-aliados-de-brasil-se-pronuncian-sobre-la-construccion-de-carreteras-en-la-amazonia/

Brondo K (2013) Land grab: green neoliberalism, gender, and Garifuna resistance in Honduras. University of Arizona Press, Tucson

Brosius PJ (1999) Green dots, pink hearts: displacing politics from the Malaysian Rain Forest. Am Anthropol 101(1):36–57

Bryant R, Bailey S (2000) Third world political ecology. Routledge, London and New York

Campbell B (2010) Rhetorical routes for development: a road project in Nepal. Contemp South Asia 18(3):267–279

Campbell JM (2012) Between the material and the figural road: the incompleteness of colonial geographies in Amazonia. Mobilities 7(4):481–500

Carvalho GO, Nepstad D, McGrath D, Diaz MDV, Santilli M, Barros AC (2002) Frontier expansion in the Amazon: balancing development and sustainability. Environ Sci Policy Sustain Dev 44(3):34–44

Chase Smith R, Salazar M (2014) La Seguridad Territorial en el Limbo: El estado de las comunidades indígenas en el Perú, Informe 2014. Instituto del Bien Común, Lima

Clech Lam M (2004) Remembering the country of their birth: Indigenous peoples and territoriality. J Int Aff 57(2):129–150

Congreso de la República (2013) Proyecto de ley que Declara de Necesidad Pública e Interés Nacional la Construcción de la Carretera de Integración Regional de Madre de Dios. Comisión de Transportes y Comunicaciones, Lima

Congreso de la República (2017) Proyecto de Ley que Declara de Necesidad Pública e Interés Nacional la Construcción de la Carretera de Integración Regional Madre de Dios, Tramo: Puerto Maldonado - Mazuco - Limonchayoc - Gamitana - Salvación. Congreso de la República del Perú

Cupples J, Prieto M, Palomino-Schalscha M (2019) Latin American development—editor's introduction. The Routledge handbook of Latin American development. Routledge, London, pp 1–11

Dalakoglou D, Harvey P (2012) Roads and anthropology: ethnographic perspectives on space, time and (im)mobility. Mobilities 7(4):459–465

Delaney D (2005) Territory: a short introduction. Blackwell Publishing, Oxford

Díaz R, Miranda J (2014) Áreas Naturales Protegidas en el Perú: Efectos sobre la Deforestación y su Relación con el Bienestar de la Población Amazónica. In: Barrantes R, Glave M (eds) Amazonía Peruana y Desarrollo Económico. GRADE, IEP, Lima, pp 209–245

Dourojeanni MJ, Barandiarán A, Dourojeanni D (2010) Amazonía Peruana en 2021 - Explotación de recursos naturales e infraestructura. Bois et Forets Des Tropiques 305(3):77–82

El Peruano (2017) Ley que Declara de Necesidad Pública e Interés Nacional la Construcción de la Carretera Iquitos - Saramiriza para su Interconexión con la Costa Norte. Diario El Peruano, p 4

Escobar A (2008) Territories of difference: place, movements, life, redes. Duke University Press, Durham and London

Figallo M, Vergara K (2014) La Amazonía Peruana de Hoy. In: Barrantes R, Glave M (eds) Amazonía Peruana y Desarrollo Económico. GRADE, IEP, Lima, pp 47–108
Gallice G, Larrea-Gallegos G, Vásquez-Rowe I (2017) The threat of road expansion in the Peruvian Amazon. Oryx, 1–9
Global Witness (2019) El justiciero forestal: Por qué se debe devolver la independencia a OSINFOR y expandir sus poderes. Global Witness, London
González GL (2012) El derecho ambiental en la integración regional de Suramérica. Letras Verdes 12:42–56
Greene S (2006) Getting over the Andes: the geo-eco-politics of Indigenous movements in Peru's twenty-first century Inca Empire. J Lat Am Stud 38:327–354
Greene S (2009) Caminos y Carretera: Acostumbrando la indigenidad en la selva peruana. IEP Instituto de Estudios Peruanos, Lima
Grosfoguel R (2007) The epistemic decolonial turn. Cult Stud 21(2–3):211–223
Gudynas E (2016a) Beyond varieties of development: disputes and alternatives. Third World Q 37(4):721–732
Gudynas E (2016b) Climate change, the quadrilemma of globalization, and other politically incorrect reactions. Globalizations 13(6):938–942
Haines S (2017) Imagining the highway: anticipating infrastructural and environmental change in Belize. Ethnos, 1–20
Harvey P, Jensen CB, Morita A (2017) Introduction: infrastructural complications. Infrastructures and social complexity: a companion. Routledge, London, pp 1–42
Harvey P, Knox H (2008) Otherwise engaged. J Cult Econ 1(1):79–92
Harvey P, Knox H (2012) The enchantments of infrastructure. Mobilities 7(4):521–536
Harvey P, Knox H (2015) Roads: an anthropology of infrastructure and expertise. Expertise, Ithaca, NY
Hecht SB (2008) The last unfinished page of genesis: Euclides da Cunha and the Amazon. Novos Cadernos NAEA 11(1):5–38
Howe C, Lockrem J, Appel H, Hackett E, Boyer D, Hall R et al (2016) Paradoxical infrastructures: ruins, retrofit, and risk. Science technology and human values 41(3):547–565
Humala O, Cateriano P (2015) Oficio 243-2015-PR. Congreso de la República, Lima
Igoe J (2004) Conservation and globalization: a study of National Parks and Indigenous communities from East Africa to South Dakota. Wadsworth/Thompson Learning, Belmont, CA
INRENA (2008) Plan Maestro de la Reserva Comunal Amarakaeri 2008–2012. Ministerio de Agricultura, Lima
Jarrier A (2012) Seminario Internacional: "Todos los caminos conducen al Brasil. La geopolítica del plan de Integración de Infraestructura Regional Suramericana - IIRSA - en la región andina." In Bulletin de l'Institut français d'études andines, vol 41, pp 303–307
Jiménez F (2017) Veinticinco años de Modernización Neocolonial: Crítica de las Políticas Neoliberales en el Perú. Instituto de Estudios Peruanos, Lima
Kanai JM (2016) The pervasiveness of neoliberal territorial design: cross-border infrastructure planning in South America since the introduction of IIRSA. Geoforum 69:160–170
Kernaghan R (2012) Furrows and walls, or the legal topography of a frontier road in Peru. Mobilities 7(4):501–520
Kohler F, Brondizio ES (2016) Considering the needs of indigenous and local populations in conservation programs. Conserv Biol 31(2):245–251
Larkin B (2013) The politics and poetics of infrastructure. Annu Rev Anthropol 42(September):327–343
Larrea-Gallegos G, Vazquez-Rowe I, Gallice G (2016) Life cycle assessment of the construction of an unpaved road in an undisturbed tropical rainforest area in the vicinity of Manu National Park, Peru. Int J Life Cycle Assess 11(17/2016):1–16
Larsen PB (2016) Derechos indígenas, gobernanza ambiental y recursos en la Amazonía peruana, Hacia una antropología de la posfrontera. Instituto de Estudios Peruanos, Lima

Leal D, Salisbury D, Faquín Fernandez J, Cauper Pezo L, Silva J (2015) Ideas cambiantes sobre territorio, recursos y redes políticas en la Amazonía indígena: un estudio de caso sobre Perú. J Lat Am Geogr 14(2):181–204
McEwan C (2009) Postcolonialism and development. Routledge, London
McLeod J (2007) Introduction. In: McLeod J (ed) The Routledge companion to postcolonial studies. Routledge, London and New York, p 2007
Mignolo WD (2000) Local histories/global designs: coloniality, subaltern knowledges, and border thinking. Princeton University Press, Princeton and Oxford
MINAM (2016) Dossier: ¿Por qué no debe aprobarse el predictamen de insistencia de la ley sobre la carretera Madre de Dios, tramo Nuevo Edén - Boca Manu - Boca Colorado?. Ministerio del Ambiente, Lima
Morel J (2014) De una a muchas Amazonías: Los Discursos sobre 'La Selva' (1963-2012). In: Barrantes R, Glave M (eds) Amazonía Peruana y Desarrollo Económico. GRADE, IEP, Lima, pp 21–46
Murray-Li T (2007) The will to improve: governmentality, development and the practice of politics. Duke University Press, Durham
Murray-Li T (2014) Land's end: capitalist relations on an Indigenous frontier. Duke University Press, Durham and London
Ohl J, Wezel A, Shepard GH, Yu DW (2008) Swidden agriculture in a protected area: the Matsigenka native communities of Manu National Park, Peru. Environ Dev Sustain 10(6):827–843
Oliart P, Biffi V (2010) Territorialidad indígena, conservación y desarrollo: Discursos sobre la biodiversidad en la Amazonía peruana. Instituto del Bien Común, Lima
Peluso DM, Alexiades MN (2005) Indigenous urbanization and Amazonia's post-traditional environmental economy. Tradit Dwell Settl Rev 16(2):7–16
Peluso NL, Lund C (2011) New frontiers of land control: introduction. J Peasant Stud 38(4):667–681
Pieck SK (2011) Beyond postdevelopment: civic responses to regional integration in the Amazon. J Cult Geogr 28(1):179–202
Pieck SK (2013) Asphalt dreams: road construction and environmental citizenship in Peru. Dev Change 44(5):1039–1063
Pieck SK (2015) "To be led differently": neoliberalism, road construction, and NGO counter-conducts in Peru. Geoforum 64:304–313
Pieck SK, Moog SA (2009) Competing entanglements in the struggle to save the Amazon: the shifting terrain of transnational civil society. Political Geogr 28:416–425
Portugal V, Linares D (2018) La Hidrovía Amazónica: Recomendaciones y retos para el cumplimiento de estándares ambientales y sociales. http://www.dar.org.pe/uncategorized/la-hidrovia-amazonica-recomendaciones-y-retos-para-el-cumplimiento-de-estandares-ambientales-y-sociales/
Quinhoes RC (2011) Redes y territorio: la iniciativa IIRSA en foco. Espacio y Desarrollo 23:5–29
Ráez Luna E (2017) Pueblos Indígenas y Conservación de la Naturaleza: ¿Debates teóricos o urgencias vitales? El caso del Parque Nacional del Manu. In: Ráez Luna E (ed) Derechos Ancestrales y Conservación de la Naturaleza en Debate: El caso de los Pueblos Indígenas del Parque Nacional del Manu. Seminario Permanente de Investigación Agraria (SEPIA), Lima
Raycraft J (2018) Circumscribing communities: marine conservation and territorialization in southeastern Tanzania. Geoforum 100(2019):128–143
Robbins P (2004) Political ecology: a critical introduction. Blackwell Publishing, Oxford
Royo-Villanova J (2017) La otra cara del baguazo. Planeta, Lima
Salisbury DS, López JB, Vela Alvarado JW (2011) Transboundary political ecology in Amazonia: history, culture, and conflicts of the borderland Asháninka. J Cult Geogr 28(1):147–177
SERNANP (2013) Parque Nacional del Manu, 40 años (SERNANP). Biblios, Lima
SERNANP (2014) Plan Maestro del Parque Nacional del Manu 2013–2018. Ministerio del Ambiente, Peru
SERNANP (2017) Oficio N°620–2017 - SERNANP - DGANP. Ministerio del Ambiente, Lima

Shepard GH, Rummenhoeller K, Ohl-Schacherer J, Yu DW (2010) Trouble in Paradise: Indigenous populations, anthropological policies, and biodiversity conservation in Manu National Park, Peru. J Sustain For 29(2–4):252–301
SPDA. (2017a). Aprueban dictamen que pone en riesgo áreas protegidas y reservas indígenas en Ucayali. http://www.actualidadambiental.pe/?p=45626
SPDA (2017b) Carretera Iquitos-Saramiriza: Viabilidad del proyecto e impactos en la Amazonía. http://www.actualidadambiental.pe/?p=46697
SPDA (2018) Purús (Ucayali): 7 de 8 candidatos proponen carreteras que afectarían áreas protegidas e indígenas aislados. http://www.actualidadambiental.pe/?p=51980
Sunderlin WD, Belcher B, Santoso L, Angelsen A, Burgers P, Nasi R, Wunder S (2005) Livelihoods, forests, and conservation in developing countries: an overview. World Dev 33(9 SPEC. ISS.):1383–1402
Taylor C (2007) Colonial empires: The Spanish and Portuguese empires. In: McLeod J (ed) The Routledge companion to postcolonial studies. Routledge, London and New York
Torres López F (2016) The highway tearing through the heart of the Amazon. Ojo Público, 1–22. http://ojo-publico.com/sites/apps/amarakaeri-parte1-english/
Tsao T (2016) Indigenous agency and compliance: contemporary literature about Dayaks. Pmla 131(3):686–700
Tsing AL (2000) Inside the economy of appearances. Public Cult 12(1):115–144
Tsing AL (2005) Friction: an ethnography of global connection. Princeton University Press, Princeton
Tsing AL (2009) How to make resources in order to destroy them (and then save them?) on the salvage frontier. In: Histories of the future, pp 51–73
Uribe S (2017) Frontier road: power, history and the everyday state in the Colombian Amazon. Wiley, Hoboken
Urrunaga J, Johnson A, Orbegoso I, Mulligan F (2012) La máquina lavadora: Cómo el Fraude y la Corrupción en el sistema de concesiones están Destruyendo el Futuro de los bosques del Perú. Environmental Investigation Agency, London, Washington DC
Urrunaga J, Johnson A, Orbegozo D (2018) El Momento de la Verdad: Oportunidad o Amenaza para la Amazonía Peruana en la lucha contra el Comercio de la Madera Ilegal. Environmental Investigation Agency, London, Washington DC
Vadjunec JM, Schmink M, Greiner AL, Vadjunec JM, Schmink M, New ALG (2011) New Amazonian geographies: emerging identities and landscapes. J Cult Geogr 3631(1):1–20
van Dijck P (2008) Troublesome construction: the rationale and risks of IIRSA. Revista Europea de Estudios Latinoamericanos y Del Caribe 85:101–120
Vergara K, Figallo M, Glave M (2014) Infraestructura en la Amazonía Peruana: una propuesta para proyectar cambios en la cobertura boscosa en la carretera Pucallpa-Cruzeiro do Sul. In: Barrantes R, Glave M (eds) Amazonía Peruana y Desarrollo Económico. GRADE, IEP, Lima, pp 161–208
Walker H (2012) Demonic trade: debt, materiality, and agency in Amazonia. J Roy Anthropol Inst 18(1):140–159
Watts MJ (2018) Frontiers: authority, precarity, and insurgency at the edge of the state. World Dev 101:477–488
West P (2006) Conservation is our government now: the politics of ecology in Papua New Guinea. Duke University Press, Durham and London
Wilson F (2004) Towards a political economy of roads: experiences from Peru. Dev Change 35(3):525–546
Zhouri A (2010) "Adverse forces" in the Brazilian Amazon: developmentalism versus environmentalism and Indigenous rights. J Environ Dev 19(3):252–273
Zibechi R (2006) IIRSA: la integración a la medida de los mercados. Ecología Política 31:19–25

Chapter 4
Ethnographic Explorations. Methodological Approach and Consideration to Explore First Roads

Abstract This chapter describes the methodological aspects of this research project, and the rationale behind how it was conducted. It first explores the qualitative ethnographic approach chosen, to then detail the methods used and participants involved. Finally, it unpacks the positionality of both researchers with regards to this project.

Keywords Ethnographic approach · Positionality · Political ecology · Road ethnography

4.1 Choosing a Qualitative Ethnographic Approach

For this research, a social constructivist epistemology was adopted. Challenging any pretensions of universal truths, a social constructivist approach focuses on understanding different social constructions of the social world (Mikkelsen 2005). In other words, this epistemological stance seeks to explore the different meanings people give to different events and processes (Denzin and Lincoln 1994), which are developed through interactions with one another and influenced by specific contexts (LeCompte and Schensul 1999).

Based on this epistemology, a qualitative ethnographic approach was chosen in order to explore multiple viewpoints and to engage with how different groups and individuals make sense of their experiences (Brockington and Sullivan 2003; Crang and Cook 2007; Denzin and Lincoln 2005). In particular, an ethnographic approach allows researchers to engage in-depth with how different people understand their world (Crang and Cook 2007). But together with exploring people's opinions and perceptions, it also addresses the political, economic, social and cultural context in which they are situated (LeCompte and Schensul 1999). Ethnography's emphasis on bringing together people's opinions and the broader context in which first roads are developed was crucial for this research. Roads, and in particular first roads, are key expressions of broader social, cultural and economic contexts and processes, while also impacting individuals and groups directly (Dalakoglou and Harvey 2012).

Furthermore, an ethnographic approach is especially useful, as we have chosen to analyse the political conflict surrounding the Manu road from a Political Ecology

E. Salazar Moreira and M. Palomino-Schalscha, *Road Expansion in the Peruvian Amazon*, SpringerBriefs in Latin American Studies, https://doi.org/10.1007/978-3-030-47182-8_4

perspective. Political Ecology studies the impacts of global political and economic systems on local landscapes and livelihoods (Robbins 2004). It does so by critically analysing how power dynamics and political processes shape the environment, instead of focusing only on environmental degradation as an effect (Bryant and Bailey 2000; Robbins 2004). Ethnography has proven very relevant to analyse processes that shape the frontier under a Political Ecology lens (Little 2007; Uribe 2017), while allowing to bridge considerations across the natural and the social sciences (Little 2007). Its capacity to 'surprise' the researcher, instead of taking them through a pre-determined path, also prompts the innovative reasonings and critical understandings essential for Political Ecology analysis (Bryant and Bailey 2000; Tsing 2005; Van Maanen 1988).

This research, grounded in the particularities of Manu, contributes to the growing number of authors around the world using ethnographic approaches to unpack the relationships, conflicts and imaginaries that evolve around roads. These authors working across a range of disciplines include, among others, Campbell (2010) and his work in Nepal, Haines (2017) in Belize, Kernaghan (2012) and Wilson (2004) in Peru, Nielsen (2012) in Mozambique and Uribe (2017) in Colombia.

4.2 How This Research Was Conducted

4.2.1 Ethnographic Fieldwork and Storytelling

In order to develop this research, Eduardo conducted two months of fieldwork in the Manu province between April and June, 2017, as part of his master's dissertation. However, this research is also informed by his previous experience living in the provincial capital of Salvación and working for an NGO that promoted sustainable livelihoods in nearby communities for two years between 2014 and 2016. Together, these experiences provide insights similar to what Tsing describes as 'patchwork ethnographic fieldwork', which capture chains of influence and power structures that affect specific issues (2005, p. x). The fieldwork involved six sites that were already connected to the road, or on-road communities, and three sites that were expecting to be connected to it soon, or off-road communities (see Table 4.1).

Following the ethnographic approach mentioned above, participant observation and semi-structured interviews were conducted. Semi-structured interviews helped develop fluid dialogue with participants, allowing for flexibility to explore their interests and opinions emerging during the conversation. Working with open-ended questions, they allowed Eduardo to promote and listen to participants' reflections, feelings and opinions (Kitchin and Tate 2000). On the other hand, participant observation provided crucial insights into the context of everyday life for those living in the area. This was facilitated by the closeness that this method allows, which often reveals meaningful details about the complex relations that roads bring about (Dalakoglou and Harvey 2012; Kernaghan 2012). By becoming immersed in the

Table 4.1 Fieldwork locations and participants

Site	Time spent on Site	On-road/Off-road	Number of participants	Type of participants
Boca Manu and Isla de los Valles*	5 days	Off-road	12	District Mayor, District Deputy Mayor, Community Chief, Community Vice-Chief, Local Student Boarding-House Staff (2), Health Centre Staff, Merchants/Shop Owners (3), Restaurant Owner, School Headmaster
Diamante	5 days	Off-road	9	Community Vice-Chief, Community Council Members (3), Primary School Headmistress, Health Centre Staff, Local fisherman, Local Tour boat Drivers (2)
Itahuanía	Passed by	On-road	1	Community Council Member
Nuevo Edén	Passed by	On-road	1	Community President
Salvación	12 days	On-road	10	Head of Manu province healthcare network, MNP Ranger, MAA Manager, Timber mill owners (2), Manu Provincial Mayor, Manu-SRG Manager, Manu-SRG Head of Infrastructure, Banana Wholesalers (2)
Santa Cruz	2 days	On-road	4	Banana wholesalers (4)

(continued)

Table 4.1 (continued)

Shintuya	4 days	On-road	5	Community Vice-Chief, Harakmbut Activist, Harakmbut Cultural Advocate, Local farmers (2)
Shipetiari	2 days	On-road	2	Community Chief, Local Farmer
Other**	–	–	2	Ministry of Culture Staff, NGO Staff
Total number of participants			46	

*Boca Manu and Isla de los Valles are adjacent sites
**Interviews conducted in Cusco and Lima

rhythms, routines and activities of the communities, Eduardo had access to experience how these experiences felt firsthand, as well as to how participants work within these circumstances and interact with each other (Crang and Cook 2007).

This combined approach allowed Eduardo to produce a set of what Van Maanen calls 'impressionist tales' (1988, p. 107), stories that provide rich accounts of what was observed, combined with the researcher's own experience. These tales, as well as some pictures taken while on the field, have been included in this book as a way of communicating Eduardo's ethnographic findings through a lively narrative that supports a deeper and more grounded understanding. Including this type of ethnographic story responds to the need for human geographers to adopt the role of storytellers when attempting to explore issues that combine politics, culture and the environment (Cameron 2012). Narrative approaches like storytelling can be more effective than other, perhaps more 'formal', approaches in capturing the causal development of issues observed (Aunger 1995), as well as explaining the production of power relations (Cameron 2012). These two aspects, causality and power, are crucial elements to analyse issues from the Political Ecology perspective (Bryant and Bailey 2000; Robbins 2004). Furthermore, Eduardo's often personal and very local stories can also be expressions of the broader social and political context in which road expansion takes place in the frontier (Cameron 2012). This type of narrative also makes academic works more accessible to the wider public, while facilitating a connection with the living immediacy of experience as well (Maggio 2014). Additionally, ethnographic stories that focus on situations of 'distress' provide a vivid narrative that has the power to reveal the failures of policies based on globalized viewpoints and offer alternatives closer to local situations (Tsing 2005). The accessibility and relatability of stories, as well as their capacity to reveal policy failures, also facilitate the political effects needed to address the issues discussed in academic writing (Cameron 2012). Thus, Eduardo's storytelling facilitates a more nuanced discussion around the case of Manu, makes this discussion more relatable and accessible and facilitates the type of change needed to address the issues this case presents.

4.2.2 *Sampling and Recruitment*

Despite the difficulty to talk to informants Eduardo was not already acquainted with or that did not hold leadership positions, he was able to interact with and interview a wide array of stakeholders in Manu. In total, 39 semi-structured interviews were conducted and audio-recorded. Interviews were conducted in Salvación, the provincial capital; the Native Communities of Shintuya, Shipetiari, Diamante and Isla de los Valles and in the settlements of Santa Cruz, Boca Manu, Itahuanía and Nuevo Edén (see Table 4.1). Verbal consent was sought with the authorities of each Native Community visited after discussing the purpose of Eduardo's presence and the methods he would apply to interact with community members. Consent to access non-Indigenous communities was not required but was nonetheless sought through Eduardo's requests to local authorities and community Presidents. All informants, including those not interviewed but with whom Eduardo interacted with regarding this research, were informed about his purposes before seeking their verbal consent and any data gathering done was notified to them. Verbal consent was chosen over written consent in order to make the process less intimidating to potential informants whose literacy skills were limited. The names of all informants have been changed for confidentiality, but the names of identifiable public figures, those of organizations and place names have been kept to provide a concrete context. All interviews were transcribed and translated by Eduardo to be later analysed.

Including participants at different social, political and administrative levels in various relevant locations allowed to approximate the analysis to the multi-scale approach that both Leal et al. and Salisbury et al. utilize in their own work on territorial issues in Peruvian Amazonia (Coombes et al., 2012 cited in Leal et al. 2015; Offen, 2004 cited in Salisbury et al. 2011). Both authors highlight the importance of this method to identify networks of influence that would remain unseen if a more specific sample of informants was the only one considered (Leal et al. 2015; Salisbury et al. 2011). This enabled the analysis presented here to penetrate the complex network of actors present in Amazonia and their engagement with various development models (Vadjunec et al. 2011).

4.2.3 *Review of Documents and Communications*

Together with interviews and participant observation, several official documents and informal communications produced by conflicting stakeholders were reviewed for this research. Together, they reveal power dynamics between the stakeholders as well as the different rationales articulated around the road. Some examples of the documents considered include official declarations by GOREMAD (2018), the 2013 Law Proposal that sought to declare the road a 'public necessity' (Congreso de la República 2013), the minutes of planning meetings between road stakeholders, social media communications by GOREMAD and the Manu SRG. Also, voices against the road were explored in official communications by the Peruvian National Government and multi-ethnic Indigenous organizations regarding road construction (AIDESEP 2018; Humala and Cateriano 2015; MINAM 2016).

4.3 Co-authorship and Positionality Matters

Despite the usefulness of narrative (Cameron 2012), we needed to discuss Eduardo's ethnographic stories in a formal language in order to weave them effectively to academic debates. Marcela's co-authorship was crucial in this aspect given her experience with similar research and contexts, even when she was not involved in this project's fieldwork. This makes it useful for us to discuss our positionality as collaborating researchers and its potential implications in this work, as well as the stylistic effects of this type of relationship on our writing.

Both of us hold somehow similar but at the same time very different positions that inform this research. Despite being Peruvian and having worked in the area before, Eduardo's positionality was seen as an urban, educated male who has represented a foreign NGO in the past, and was then conducting postgraduate studies in the UK involving this research project. In fact, in the field, some participants mentioned his position of privilege, especially regarding his urban background and his access to prearranged means of transportation through his work connections. However, Eduardo's experience with local livelihoods, culture and daily life, as well as his familiarity with perspectives regarding the issues analysed, enabled him to establish rapport and relate with most participants in an open manner as well as to grasp nuances of the context. Thus, as a whole, Eduardo's positionality was of both being an insider and outsider. A certain degree of 'insider-ness' facilitated this research (Eppley 2006), while the remaining 'outsider-ness' had to be negotiated on an ongoing basis.

But Eduardo's privileged access to transportation not only influenced his direct interaction with participants. More generally, it allowed him to avoid the spatial, seasonal and security biases described by Chambers (2008). By travelling to rural communities only accessible by river, he could avoid the 'urban', 'tarmac' and 'roadside biases', which this author describes as limitations to learning about rural peripheries. His arrival during a deferred rainy season on an El Niño year played a noteworthy role in the avoidance of 'seasonal bias' (Chambers 2008). The heavy rains delayed travels along the Manu Road, but this aided his observation of the difficulties that are present in travel along this infrastructure. On the other hand, the level of water on the AMD river made travel along it relatively easier. Eduardo also avoided 'security bias' by travelling to places where transport hazards and other physical dangers were inevitable, especially considering the contemptuous nature of this research topic and the lack of police in the area (Chambers 2008). He was, however, unable to avoid 'elite bias' and 'people bias' as the short length of his stay seldom allowed him to talk in-depth to people outside community leadership and informants he was already acquainted with (Chambers 2008).

On the other hand, Marcela was not involved in the conduction of fieldwork. Being originally from Chile, she also has an urban background. She has experience working with students doing research in the Amazon, as well as conducting research herself on a first road within Mapuche communities in Chile. Although in the insider-outsider continuum she is definitively an outsider to this research, she is to some extent familiar with the nuances of this research case study due to her previous experiences and expertise on Development Studies and Indigenous Geographies.

In this way, Eduardo's firsthand experience in Manu makes him the ideal 'storyteller' for this work, allowing him to address the need for human geographers to assume this role (Cameron 2012). At the same time, Marcela's role as an academic advisor has shaped the discussions we both conduct in this publication. This, however, introduces some complexity to our point of view when writing, requiring us to switch between perspectives along this work. As in this sentence, we use the first-person plural ('we') when discussing issues as co-authors. We also use the first-person singular ('I') in Eduardo's ethnographic accounts as a way of conveying his immersed point of view directly to the reader. Finally, we also use the third person ('Eduardo', 'Marcela', 'he', 'she') when referring to either of us from our joint perspective as co-authors. Clarifying this necessary complexity will allow us to tell stories that engage the wider public from Eduardo's perspective, while also contributing effectively to relevant academic debates as co-authors.

This research was approved by the Research Ethics and Integrity Committee of the School of Geosciences of the University of Edinburgh, represented by Dr. S.J. Shackley, Chair of the Committee, on the 30th of March, 2017.

References

AIDESEP (2018) Pronunciamiento de AIDESEP sobre la LEY N° 30723. AIDESEP Website (30723):1–2. http://www.aidesep.org.pe/pronunciamiento-de-aidesep-sobre-la-ley-no-30723/
Aunger R (1995) On ethnography: storytelling or science? Curr Anthropol 36(I):97–130
Brockington D, Sullivan S (2003) Qualitative research. In development fieldwork: a practical guide. Sage, London
Bryant R, Bailey S (2000) Third world political ecology. Routledge, London and New York
Cameron E (2012) New geographies of story and storytelling. Prog Hum Geogr 36(5):573–592
Campbell B (2010) Rhetorical routes for development: a road project in Nepal. Contemp South Asia 18(3):267–279
Chambers R (2008) Rural development tourism: poverty unperceived. In Revolutions in development inquiry published. Earthscan, pp 25–48
Congreso de la República (2013) Proyecto de ley que Declara de Necesidad Pública e Interés Nacional la Construcción de la Carretera de Integración Regional de Madre de Dios. Comisión de Transportes y Comunicaciones, Lima
Crang M, Cook I (2007) Doing ethnographies. Sage, London, Thousand Oaks, New Delhi, Singapore
Dalakoglou D, Harvey P (2012) Roads and anthropology: ethnographic perspectives on space, time and (im)mobility. Mobilities 7(4):459–465
Denzin NK, Lincoln YS (1994) Handbook of qualitative research. Sage, Thousand Oaks
Denzin N, Lincoln Y (2005) Introduction: the discipline and practice of qualitative research. The Sage handbook of qualitative research. Sage, Thousand Oaks, pp 1–32

Eppley K (2006) Defying insider-outsider categorization: one researcher's fluid and complicated positioning on the insider-outsider continuum. Forum Qual Soc Res 7(3)

GOREMAD (2018) Pacto por el Desarrollo Económico-Social y por un Madre de Dios sin Pobreza ni Discriminación. Puerto Maldonado, Madre de Dios

Haines S (2017) Imagining the highway: anticipating infrastructural and environmental change in Belize. Ethnos, 1–20

Humala O, Cateriano P (2015) Oficio 243-2015-PR. Congreso de la República, Lima

Kernaghan R (2012) Furrows and walls, or the legal topography of a frontier road in Peru. Mobilities 7(4):501–520

Kitchin R, Tate NJ (2000) Conducting research into human geography. Theory, methodology and practice. Harlow: Pearson Education

Leal D, Salisbury D, Faquín Fernandez J, Cauper Pezo L, Silva J (2015) Ideas cambiantes sobre territorio, recursos y redes políticas en la Amazonía indígena: un estudio de caso sobre Perú. J Lat Am Geogr 14(2):181–204

LeCompte MD, Schensul JJ (1999) Design and conduct ethnographic research. Sage, London

Little PE (2007) Political ecology as ethnography: a theoretical and methodological guide. Horizontes Antropológicos, 3

Maggio R (2014) The anthropology of storytelling and the storytelling of anthropology. J Comp Res Anthropol Sociol 5(2):89–106

Mikkelsen B (2005) Methods for development work and research: a new guide for practitioners, 2nd edn. Sage, New Delhi

MINAM (2016) Dossier: ¿Por qué no debe aprobarse el predictamen de insistencia de la ley sobre la carretera Madre de Dios, tramo Nuevo Edén - Boca Manu - Boca Colorado? Ministerio del Ambiente, Lima

Nielsen M (2012) Roadside inventions: making time and money work at a road construction site in Mozambique. Mobilities 7(4):467–480

Robbins P (2004) Political ecology: a critical introduction. Blackwell Publishing, Oxford

Salisbury DS, López JB, Vela Alvarado JW (2011) Transboundary political ecology in Amazonia: history, culture, and conflicts of the borderland Asháninka. J Cult Geogr 28(1):147–177

Tsing AL (2005) Friction: an ethnography of global connection. Princeton University Press, Princeton

Uribe S (2017) Frontier road: power, history and the everyday state in the Colombian Amazon. Wiley, Hoboken

Vadjunec JM, Schmink M, Greiner AL, Vadjunec JM, Schmink M, New ALG (2011) New Amazonian geographies: emerging identities and landscapes. J Cult Geogr 3631(1):1–20

Van Maanen J (1988) Tales of the field, on writing ethnography. University of Chicago Press, Chicago

Wilson F (2004) Towards a political economy of roads: experiences from Peru. Dev Change 35(3):525–546

Chapter 5
The 'Enchantments' of Speed and Political Integration

Abstract In this chapter we analyse the way in which the 'enchantments' of speed and political integration manifest regarding the ongoing extension of the Manu Road (Harvey and Knox in Mobilities 7(4):521–536, 2012). To do so we present accounts based on the testimonies and observations gathered by Eduardo in the field, which allow us to compare the aspirations held in off-road communities with the realizations of nearby on-road communities. Our analysis reveals situations that generate overwhelming aspirations for an increase in the speed of travel and for easier contact with government instances in off-road communities. More importantly, it also reveals that these aspirations have not materialized in on-road communities. We argue that this mismatch is a manifestation of the 'enchantments' of the Manu Road, which, despite not fulfilling what communities aspire to, continues to foster support for its construction. In the following sections we take a look at each of these 'enchantments' by setting the scene with relevant stories, and the corresponding perspectives of off-road and on-road communities, which are complemented with further accounts from Eduardo's fieldwork.

Keywords Manu road · First roads · Frontier infrastructure · Enchantments · Speed · Political integration · Political ecology

5.1 Stories of the 'Enchantment' of Speed

The promise of speed is one of the main ways new roads 'enchant' (Harvey and Knox 2012). The following accounts describe Eduardo's experiences regarding the lack of speed in travel experienced in Manu, both while travelling on the AMD river and on the Manu Road. They illustrate the importance of speed in the area, and how a lack of speed persists in Manu despite the construction of a road, issues further explored in the following section.

E. Salazar Moreira and M. Palomino-Schalscha, *Road Expansion in the Peruvian Amazon*, SpringerBriefs in Latin American Studies, https://doi.org/10.1007/978-3-030-47182-8_5

5.1.1 *'Will I go? How will I go? When will I go?'—Eduardo's River Transportation Experience*

I had been lucky to get picked up from Boca Manu at all. Robert, the boat driver who had dropped me off there on his way into the National Park with a group of tourists, had forgotten about me. Melquiades, his deckhand, wasn't sure if I had asked them to pick me up or not, so he decided not to mention it. I would learn about this later. They had told me they would pass Boca Manu early in the morning of that day, on their way back up the river after leaving the tour group in Boca Colorado. '5 am, or 7 at the latest! We need to get the boat back to the lodge by that afternoon!', one of them had assured me. I had my stuff ready at 4:30 am, the day's work had already started for many in the community and chainsaws were buzzing at the timber mills.

My long wait at the port had become hopeless after 8 hours. I was so desperate not to miss them and hopeful that I would not that I kept going back to the port after my short breaks, which consisted of sitting somewhere else re-reading field notes, making new ones. After seeing me there for so long, some of my informants started approaching the stump I was sitting on, trying to comfort me, albeit in their own way. Silvia, one of the three shop owners, explained how the road foreboded the end of these long, uncertain, waits. 'With the road we wouldn't need to worry about, "Will I go? How will I go? When will I go?" You might have to wait until tomorrow. They might have missed you too. If they haven't passed Boca Manu by 3 then they won't be coming. That's too late for them to get back to Reinalandia [the lodge where they worked at]'.

The shop owners had started to like me by then, I guess their motherly instincts kicked in. 'Sometimes we wait for days to get on a boat to travel. Although you might have an easier time convincing the tour boats to pick you up because you're a gringo', said María. I asked her to confirm Silvia's theory about the 3 pm deadline. She did, it was 2 already. Then she added what she believed was the reason for my dilemma; 'The road would change so much but they don't want us to build it.'

She was referring to the SERNANP and the Ministry of the Environment (MINAM), who had taken legal action against GOREMAD to stop the unauthorized construction of the road two years earlier. The long-standing resentment against these central government entities had only grown since the controversy about the road started.

By 4 pm I was already giving up, but my familiarity with how time estimations work in rural Perú kept me coming back to my port stump. My ears were tuned to the sound of incoming boat motors; I could already distinguish them from the more permanent buzz of chainsaws. I had attempted to wave and whistle-in the few boats that had passed the town going up river since 3, they had ignored me. At 4:30 I gave up, there was no way Robert and Melquiades would make it anywhere before sundown. Going up river took much longer. 'Something must have happened', I thought, there were endless possibilities for mishaps.

As I was finishing my walk around Boca Manu, equivalent to a walk around a city block, I heard the buzz that I had been listening out for almost 12 hours. 'Will this be my chance?' I thought. 'No way but let's check it out.' I ran back to the port and, despite that the light was already dwindling, I distinguished the dark green hull and what seemed to be the familiar white logo of the voluntourism NGO I had worked for before. They were driving fast on the other side of the river. I was naively expecting them to turn into the port at any time. One of the men who worked for the shops was also expecting a boat, 'Reinalandia, right? That's them, you'd better call them. They're going full on.' My waving and whistling didn't work but his, more accustomed to the circumstances, did. They started turning into the port.

'Get on, quick! We thought you weren't coming!', was the first thing Robert told me. I was too out of breath from running back for my stuff to answer. I was introduced by Melquiades to several people hitching a ride. 'We'll only make it to Diamante. I can introduce you to some people there', offering his help was Robert's way of compensating almost having left me in Boca Manu. I felt more than compensated and was happy to accept.

I snapped out of my happiness at having easier access to another off-road community when I realized how dark it was getting. We wouldn't make it with enough light to see where we were going. Preparing for the worst, I put my recorder, camera and notepad inside a zip-bag and tried to relax looking at the amazing colours of the Amazonian sundown.

At around half past six we were able to see the few dim, generator-powered, streetlights of Diamante in the distance. That and the stars, my eyes could not pick up much else. I hoped Melquiades's eyes could, he was sitting at the bow as the bottom of the boat bumped lightly against the invisible rocks. Robert kept the boat as slow as he could, following Melquiades's instructions while avoiding getting swept by the current. I trusted Robert's experience and his instinct, as well as the stereotype about Yine people being inborn river navigators. We finally docked at the closest port in Diamante and I was able to breathe normally again.

As we walked up the trail from the port, Robert was telling me about Carmen, the community Vice-chief, who he would introduce me to. He would also introduce me to Celia, who owned the shop, the payphones and the only lodging in the community. Robert's introductions would hopefully kill several birds with one stone. As we kept walking, someone approached Robert, 'How will you get through? There's an arm that filled up this afternoon.' The rainy season had prolonged into May and what was usually a small creek through the community settlement had overflown and turned into an ample arm of the AMD river. It was uncrossable on foot, especially in the dark. 'We'll have to get back on the boat and dock in the other port. My house is that way as well', said Robert confidently. The river adventure would not be over before another boat ride in the dark.

5.1.2 'Slowly but Surely'—Travelling on the Manu Road

I had to take Mr. Cabello's bus to get to Shintuya, the first on-road community I visited. The man was cherished by everyone in Manu as he provided the only affordable transport up and down the road. He would leave Salvación at 5 am and drive his bus to Shintuya, where he would arrive at 9 am. He would drive back after as much as it took for his returning passengers to load up their belongings and cargo. He would be back in Salvación by mid-day. He had done this, with his wife as a fare collector, every single day since 2004. 'Slowly but surely' was his motto, which was reinforced by a poster of the sacred heart of Jesus that was set behind his seat, facing his passengers. He would spend his afternoons under his bus, checking that everything was in order for the next day.

I had ridden in early and been able to take the seat next to him but the overbearing sound of the motor and the creaking of the bus's moving and non-moving parts made it hard to hear him. 'It's like a *picuro* trail.', said Cabello, referring to a forest rodent. The sides of the road were overgrown with bushes that made it impossible to see much further ahead. The road's muddy surface had the shape of a cursive 'm', as the wheels of passing timber trucks had compressed the material. Water accumulated in the deeper parts, further eroding the road. 'It's like this all the way to Shipetiari. It's as if it had been forgotten', he added. As he was attempting to explain the different techniques he used to pass bridgeless rivers and other types of obstacles, he had to brake suddenly, making some of the sacks that were piled in the aisle roll forward. We had almost run into a motorcycle head on. 'It's worse if it's a car', he said, as he showed an ample smile to the motorcyclist, who returned the gesture and waved.

Other passengers overheard my curiosity about the road surface and added their own observations. Mrs. Imelda, an Andean lady whose family I had worked with before, said, 'The road is *monte.*[1] My son tells me it's scary for him to drive his motorbike on it because he can't see what's coming.' Damián, a local farmer who was transporting several sacks of chicken feed, told me that, 'Sometimes cars have to reverse five hundred metres or so to make way for incoming traffic. If you're on a slope, it's really dangerous to do that.' Cabello nodded. It was obvious that the road was not receiving the maintenance it required. This, according to Dagoberto Coaquira, head engineer of infrastructure at the SRG, needed to be frequent, as most of the road had been built employing a rudimentary technique that used the plant matter chopped down to make way for the road as a base material. The four-hour ride was as bumpy and slow as it had always been. The smile on Cabello's face, even after the umpteenth time he got out of the car onto the pouring rain to bring cargo down from the rack up top, was also as wide as it had always been.

5.1.3 *'Our Parents Didn't Even Know What a Road Was!'—The Effects of Decades of Connectivity*

Shintuya was the final stop on Mr. Cabello's bus route, which made this Harakmbut community a frequent passing place for people from off-road communities on their way to Salvación and Cusco. I had made a few friends in Shintuya during my time

[1] Overgrown.

working in Manu[2] and my relatively frequent visits had shown me that life was not much better there than in the off-road communities I had recently returned from. Regardless of being connected to the road since the early 1970s and having the presence of generous benefactors, like the Dominican Mission of San Miguel de Shintuya and Hunt Oil, or maybe because of them, life was still tough.

The changes I saw there did not reflect the rapid and significant development that was expected of road connectivity by off-road communities. Improvements to urban, school and health infrastructure, water supply and sanitation only started taking place in the late 2000's with the arrival of Hunt Oil's drilling exploration crew (INRENA 2008). And these enhancements seemed far from ideal to locals. I interviewed Edmundo Kendero, a local farmer and former chainsaw operator, at his house near the communal port. Unlike those in off-road communities, he had a large padlock hanging from the door. He told me that, 'Nothing changed until recently. Our streets were improved, that's about it.' Water and drainage were available in Shintuyan households, but this had been made possible on the SRG's terms: every house had had an outhouse installed but they had been built with toilets instead of the more common latrines. 'They even want to tell us how we have to take a crap! Before we used to go in the bush and the dung beetles would do the rest', said Billy, a Harakmbut artist, craftsman and tour operator dedicated to recovering his people's traditions. Most people also complained that the sewage went straight into the Shintuya stream, which was used recreationally before. I didn't learn that until after my first swim there.

The banana market and economic connectivity hadn't facilitated many changes either. Ignacio Maqueri, whom I met at the port in Shintuya, said that 'things have improved with farming. Not a lot but a little bit. Maybe in 1% (sic). Now people can buy some rice, some sugar.' Young Shintuyans had access to new job opportunities but most were either working for illegal mining, logging or, if they were lucky, working precariously in cities like Cusco or Puerto Maldonado. Most of them left and never came back to contribute to their communities.

Other expected changes seemed like they would never come. Back in Boca Manu, Aníbal Choque, deputy mayor of Fitzcarrald, told me, 'Once the road arrives we will automatically have 24 hour energy. The power lines are set all the way up to Nuevo Edén so their extension will happen simultaneously with road construction.' Edmundo laughed when I mentioned them. 'People hope for electricity, for cell phones, for the internet. But, even with those wires up, I can't even think of when all that'll get here! We don't have any power in Shintuya, much less cell phones.' Despite the great efforts to install the cables and posts, which included clearing several metres of forest on one side of the road, the SRG had not foreseen that the power supply available to Salvación was not enough to supply communities further down the road. The provincial capital depended on energy generated in Cusco, which was reluctantly shared by that region, and was fragile and inconsistent. To overcome this insufficiency, an electrical substation had been built near the town. Yet, the

[2]Prior to his Master's dissertation work, Eduardo lived and worked in the area for 2 years. See Chap. 4 for more details.

calculations about the water flow that was supposed to generate the electricity to feed the station were mistaken. A large portion of the small funding that Manu had available had been invested in a useless power station that was not able to feed the tens of kilometres of power lines along the road. So, the cables were 'only good for the vines that grow on them', according to a woman I met on the local bus. All the communities past Salvación were still using diesel generators to have electricity. As in off-road communities, this only allowed for two hours of power every evening. 'Just enough to have dinner and for our children to do some quick studying', as a young mother in Diamante put it. Things were far from the idea people in off-road communities had for when the road reached them.

As if the scarcity of positive effects was not enough, it seemed the road had caused more harm than good. 'The loggers have looted our resources. They invaded all our territory without our consent. They just came in, no permits or anything and knocked it all down. We still don't have good trees and the few ones that we have left are far away. That's affected us a lot. Only a few people noticed, quite late, that we shouldn't have knocked down all our hardwoods', is what Edmundo said about what had happened as a consequence of road connectivity and the promotion of timber industry by the Catholic missionaries and the local government. What seemed to affect him the most was the fact that he did not have a boat. 'There are no hardwoods for me to build my boat with. Outsiders have taken all the good trees from our territory and the little profit locals made was spent in alcohol, which the loggers themselves brought for their Harakmbut workers.' Adding to this, loggers also took advantage of the Harakmbut's lack of experience with money to extract as much timber as possible for very low prices.

Edmundo's wife Soledad delved deeper into the social costs of road connectivity, 'Our parents didn't even know what a road was!' Indeed, the road must have been a shock for the Harakmbut, most of whom had lived as nomads in the forest until about a decade before the road was built. She continued, 'The people who came here called us lazy. They insulted us because we didn't do much agricultural work. We only hunted, fished and gathered fruits from the forest, as we had done before. We felt forced to leave that and plow our land. All because, to them, money was first.' Their traditional livelihoods were not the only thing incomers censured, as most interviewees also spoke of the shame they felt to speak the Harakmbut language in public.

Furthermore, according to Soledad, the change in the ecosystem brought by years of exploitation of Harakmbut territory had a cultural effect. 'It affected three things we depended on: fish, animals to hunt, and forest to gather in. Without fish we need to buy tuna cans, without animals to hunt we need to buy chicken, without fruits from the forest we need to buy sugar and rice. We need to eat to remain healthy but now we depend on having money be able to do so.' The pressure to adopt a foreign culture also fostered the adoption of vices that had become destructive for Shintuyans, 'Outsiders first brought alcohol, something we only used once a year by making *masato* for our *Sine* festival. They brought alcohol and prostitution. People from Shintuya soon began to work just to get money for alcohol and prostitutes.' Soledad continued explaining what she and other Harakmbut women had gone through when they lived with their Andean ex-husbands. To gain open access to timber in the community's territory, it became common for them to marry young Harakmbut women. By the time the timber was depleted, it became unusual to see Andean migrants living in Shintuya.

5.2 Not so Fast—Analysing Aspirations and Realizations Regarding Speed

Based on the insights provided in the two stories above, in this section we explore the aspirations of increased speed in travel held by people in off-road communities, to then examine those of people in on-road communities. Comparing these perspectives reveals the power of the Manu Road to 'enchant'.

Road construction promises speed in the form of faster, safer and more affordable travel, which is expected to bring further benefits (Harvey and Knox 2012; Kernaghan 2012; Tsing 2009; Wilson 2004). For off-road communities that depend on riverboats as their only means of transportation, roads promise to end what Harvey and Knox call 'the tyranny of delay', which severely constrains livelihoods and, as will be seen for the case of Diamante in Sect. 5.3.1, lives themselves (2012, p. 524). The aspiration for increased speed in these communities motivates dreams of a better future. However, the speed provided by roads means different things to different people, depending on their power, sources of livelihood and wealth (Tsing 2005). In the case of the Manu Road, the situation in on-road communities reveals how much more is needed to reach the levels of speed required to fulfil dreams of connectivity, which perpetuates the 'enchantment' of speed.

5.2.1 Quick, Safe and Affordable Travel—The Promise of Speed

In the absence of landslides, travelling from Salvación to Cusco takes close to ten hours. Mini-buses leave at around noon and arrive late in the evening. Adding to this are three hours travelling on Mr. Cabello's bus, which runs once a day, as well as several hours of boat travel. Considering that mini-bus and bus services run once a day, and that boats that will actually pick someone up are unpredictable, there is plenty of waiting time in between. The Law Proposal that sought to declare the Manu Road as a public necessity in 2013 attempts to present a compelling case regarding these difficulties, which SRG officials and the people of Manu have to go through to travel within and outside the province (Congreso de la República 2013). This reflects the concerns of the people in off-road communities regarding the uncertainty and delays they face when travelling. In the account in Sect. 5.1.1, Eduardo presents his own experience of uncertainty and delay. In it, the people of Boca Manu share their own stories and warnings about what he was facing during his fieldwork, which was an easy, one-off, version of what they constantly experience. As will be seen later, the situation in Diamante is much more complex in this regard, considering that Boca Manu is the capital of the Fitzcarrald district and a usual stop for tour boats on their way to the MNP. Robert, the tour boat driver who worked for Reinalandia, told Eduardo that *Diamantinos* 'normally need to wait for boats for at least two or three days'. Eduardo's account also shows how the great majority of people he spoke to firmly believed that the road would improve this situation. This belief is seen in similar cases in Peru and elsewhere (Campbell 2010; Harvey and Knox 2012; Kernaghan 2012; Tsing 2009; Wilson 2004).

Moreover, the appearance of increased speed in road travel also announces relative safety and affordability compared to riverboat travel. Eduardo's boat trip from Boca Manu to Diamante on a moonless night was a powerful example of the situations that motivate aspirations for safe travel. So did the stories of babies and small children with '*susto*'[3] after capsizing on boats with their families, or accounts of boatloads of valuable goods lost to the choppy waters of the AMD river. These and other experiences reveal that the people of Manu, like frontier dwellers around the world, have a high tolerance for risk but reducing the chances of losing their lives is something they will always seek (Harvey and Knox 2015; Murray-Li 2014; Tsing 2005). On the other hand, affordability is a key issue for these low-income communities whose members cannot afford hundreds of soles in boat fuel, or even a share of that. Luis Manuel, a restaurant owner from Boca Manu, gave an example that illustrated the change that was expected, 'Road travel from Cusco to Nuevo Edén costs 20 soles[4] on the back of a trailer. That's close to twelve hours on the road if it's done quickly. On the other hand, travel from Nuevo Edén to Boca Manu [about 40 km] is 40 soles on a cargo boat. The difference is abysmal!'

The risk and high costs of river travel slowed down public servants as well. *Profesor* Fernando Paniagua, the headmaster at the school in Boca Manu, explained how the danger and expenses of boat transportation kept teachers away from his school. Additionally, the stories that a nurse technician from Diamante told Eduardo, of a pregnant woman who did not travel for routine tests and an injured man who cannot travel for an urgent operation, show how these costs can threaten lives as well. Furthermore, situations like the death of a child at an ill-equipped health post, which Eduardo witnessed and seemed altogether common in off-road communities, made the sheer possibility of evacuating patients by road, regardless of the time of day, seem like a benefit worth any cost. Thus, in addition to faster travel, its apparent predictability, safety and lower cost, the speed provided by roads is also seen as the basis for other elements of modernity, facilitating the transport of services and goods (Harvey and Knox 2012; Pieck 2011; Wilson 2004). This is clear in the perceptions of the people of Diamante, who believed the appalling situation their public services

[3]Literally 'fright' or 'horror', a state of spiritual, mental and/or physical insufficiency caused by a traumatizing experience that requires traditional methods of healing. In Manu people tend to take their children with *susto* to tobacco healers like Soledad, our participant from Shintuya.

[4]About USD$ 6.00 at the time.

were in would be solved by state officials arriving more easily through the road. As will be seen in further chapters, the farmers in Isla de los Valles also believed that the lack of economic opportunities in their community was as an enclosure inescapable except through a road. The dire situations presented in these stories illustrate the kind of aspirations that the 'enchantments' of roads are established upon, which would make it hard to directly question the strong desire for road connectivity that the people in off-road communities have (Harvey and Knox 2008, Bennett 2001 cited in 2012; Uribe 2017).

5.2.2 *A Dead Stop—The Results of Road Connectivity*

Despite the aspirations held in off-road communities, what Eduardo saw in on-road communities was only a slight improvement. As in many other cases, the fulfilment of what the road promised was illusory (Harvey and Knox 2012; Kernaghan 2012; Wilson 2004). One of the elements of the gap between the aspirations held by off-road communities and the realizations of on-road communities is the lack of surface quality and required maintenance of the Manu Road. As in other cases, its material state significantly curtails the expected effects of its construction (Harvey and Knox 2012; Kernaghan 2012; Pieck 2011). Its fragility, its vulnerability to how it is used and to the environmental conditions affecting it, clearly interrupted the smooth flow of goods and people (Harvey et al. 2017). These effects are common throughout Peru, especially in the Amazonian regions (Barrantes et al. 2014; Larsen 2016).

This is noticeable in Eduardo's experience and that of Mr. Cabello's other bus passengers, described in Sect. 5.1.2. His passengers worry strongly about the lack of maintenance, also mentioned in studies about this road (Larrea-Gallegos et al. 2016), which prevents it from delivering the speed expected by off-road communities and even generates new risks, like vehicular accidents. Another one of Eduardo's experiences on the road was a clear example of this. He says:

> While waiting in Puerto Shipetiari to travel back to Salvación, I suddenly noticed that no cars or trailers were coming or going. One of the shop owners was talking loudly at her radio. Apparently, a large section of the road on a steep mountainside had collapsed under a trailer and the vehicle was stuck, blocking the way through for other vehicles (see Fig. 5.1). All traffic up and down the road was stopped and the machines that could fix the road could only get there the next day. After the news was announced, several men drove away swiftly on their motorcycles. They would help unload the trailers that were stuck so their cargo could be reloaded onto trailers at the other side of the obstruction. Those of us at the port would have to wait for transports that could take us to the relay point or walk to it. It seemed that frustratingly long and uncertain waits were not only part of river travel, but often remained with the construction of the road. Thus, the fragility of the road perpetuated the risks, uncertainty and delays that characterized river travel. Other travellers and I were able to make our way back to the relay point thanks to a SERNANP pickup truck that had brought Park officials to the port after doing the relay themselves. We had to walk past the landslide and wait for a trailer to drive us back to Salvación on the other side. We would arrive there late in the afternoon. The lack of speed I experienced on the road was similar to the one I went through when travelling by river.

Fig. 5.1 The road to Shipetiari collapses under a loaded trailer

Furthermore, the fragility of the Manu Road is also seasonal, increasing in the rainy season, when the water level in the tens of bridgeless rivers and creeks rises dramatically. As a banana wholesaler Eduardo spoke to explained, 'You can only travel during the dry season. When it rains the road past Shintuya is too risky to pass because of the landslides in the Muyuna [mountain]. The Mochina and Serjali [rivers] become death traps.' These expert drivers only risk travelling past this Harakmbut community when the bananas they want to buy are scarce, the rains are mild and the rivers are low. Moreover, this fragility and the dangers it generates are not only restricted to the newer stretches of this road but extend all the way into the higher areas of the cloud forest of Cusco. The curvy and muddy slopes leading down from the highlands had been the site of Eduardo's closest brush with death a few years before his fieldwork: the speedy mini bus he was on could not avoid an incoming banana trailer on the narrow road and flew off its edge. Luckily, the mini bus tumbled onto a patch of bamboo, which caught onto the luggage rack on top of the vehicle. A group of teachers travelling back from a school competition a few months before Eduardo's research visit had not been equally lucky. Thus, the issue was not exclusive to the Manu Road. All the long way to the Andean highlands, the muddy road to Cusco was unsafe given its mediocre surface quality and lack of maintenance.

What is more, the risk of travelling the Manu Road kept the monetary cost of transport unaffordable to most. The testimonies of informants from Shipetiari speak of the frustration the Matsiguenka people of this community had regarding this issue. The personal fare on a pickup truck from Salvación to Shipetiari was more than four times the fare Mr. Cabello would charge to take someone to Shintuya on his bus, even when the distance travelled was no more than double. The new road was making travel less expensive but had not made it affordable yet. Furthermore, the increased risk not only affected personal transport fares but the cost assumed by banana wholesalers as well. The testimonies of banana wholesalers, detailed in the following chapter, explain how the dangers these merchants faced along the journey from the Andean highlands to the lowlands of Manu made commerce with the farthest on-road communities unviable. The consequences of a lack of speed went beyond the promise of speed itself (Harvey and Knox 2012; Pieck 2011; Wilson 2004).

Despite these high costs, Chief Magno Abelardo did sound satisfied with the possibility of evacuating patients with medical emergencies from Shipetiari. The slight increase in speed provided by the Manu Road seemed valuable to overcome life-or-death situations like the ones the Yine people of Diamante still faced. This sole benefit seemed justification enough to build the road but the destination of the evacuees, the still precarious health posts in Shintuya or Salvación, did not promise much to them either. Most patients would still have to endure the risky trip to the city of Cusco, ten hours long in good conditions, for the effective treatment of most emergencies.

Regardless of how the little speed provided by the Manu Road would not satisfy most of the aspirations of on-road communities, it still seemed to be valuable for incomers. The way in which new roads change the framework of incentives in the frontier, creating the perception of increased access to an 'empty' resource-rich area, attracts migrants who are willing to face the risks still present on these infrastructures as a last resort (Barbier 2012; Barrantes and Glave 2014; Pieck 2011; Salisbury et al. 2011; Tsing 2005; Vadjunec et al. 2011). The likely surge of incomers in search of any economic opportunities that emerge tends to generate confrontations with already existing populations, often resulting in their dispossession (Barbier 2012; Leal et al. 2015; Pieck 2013; Tsing 2005). This is clearly the case in the Peruvian Amazon and in Madre de Dios especially, as it is the Amazonian region with the highest rate of Andean migrants, as well as of recent deforestation (Barrantes and Glave 2014; Figallo and Vergara 2014; Leal et al. 2015; Oliart and Biffi 2010; SPDA 2017).

In the account presented in Sect. 5.1.3, the effects that the Harakmbut people of Shintuya went through after decades of road connectivity are clear. Westernized Andean and urban migrants discriminated against them because of their livelihoods, cultural practices and language. Incoming loggers abused the resources in their territories, affecting their way of life, and abused their people, affecting their social dynamics. The effects of road connectivity seem to be similarly pernicious decades later, as will be seen in the account of Eduardo's experience in Shipetiari, presented in Sect. 5.3.2. The fear that the people of this Matsiguenka community have of becoming involved in the illegal trade taking place near their community is a clear example of this. The increase in timber sales, although currently seen by them as a positive, bears the serious risks of depletion that Shintuyans warned about.

Thus, as the speed provided by the muddy, ill-maintained and bridgeless Manu Road was minimal, it provided little benefit to the communities connected to it, except in desperate situations. The absence of the expected speed in on-road communities jeopardized the fulfilment of the promises the road carried but was not being considered in off-road communities, a manifestation of the 'enchantment' of speed. What is more, the 'enchantments' of the Manu Road were so powerful that the risks of a surge in migration, notorious in on-road communities, were also being ignored. Having revealed this, we now move on to a similar analysis regarding the 'enchantments' of political integration (Harvey and Knox 2012). The results, not surprisingly, are of a similar nature.

5.3 Stories of the 'Enchantment' of Political Integration

Off-road communities aspire to have an increased presence of the state in the area for a number of reasons, including alleged better access to public services. These aspirations reinforce support for the road, which is seen as crucial to improve their current situation. In this section, we include two accounts from Eduardo's fieldtrip, which explore these issues.

5.3.1 '... Just the Latrines'—The Dire Lack of Public Services in Diamante

After our ride in the dark (see Sect. 5.1.1), we arrived in Diamante's main port and went up the long stairs. Everyone went their own way except Robert, who took me along the town's main walkway to Celia's shop. It had completely slipped my mind that I had arrived on Mother's Day weekend. The shop was full of different-sized plastic buckets filled with packaged foods, and wrapped in cellophane with signs that said, 'Happy Mother's Day!' Some Yine women were putting some more together with stuff from inside the shop's cellar. There Robert introduced me to Celia, the shop owner, and Carmen, the Vice-chief, both of whom I noticed to be of Andean origin. I explained my purpose in the community to them. Carmen seemed interested in telling me her point of view about the road but said she and everyone in the community would be too busy organizing the weekend's festivities. She kept on giving orders to Celia and the troupe as I shot a few questions about the celebrations at her.

Intending to introduce myself to him as well, I asked Carmen about Jorge Ricardo, a Yine man whom I had been told was the Chief by most informants in Salvación, the provincial capital, and elsewhere. She was very emphatic when saying, 'I am the Chief now! Jorge is already living in Puerto Maldonado [the regional capital] because of his work for COHARYIMA.[5]' She seemed to have taken the lead of communal organization for the celebrations and was quick to accept my presence with a few conditions. 'First you must meet the community at our gathering, that will make things easier for everyone. You must let me know who you will be interviewing beforehand. And you would be better considered if you donate at least two Mother's

[5] *Consejo Harakmbut, Yine y Machiguenga*, a multi-ethnic Indigenous organization that represents the interests of the communities of these three ethnic groups in the Manu province.

Day baskets. Please choose your donations from the ones we have here.' I chose two, a green and a blue bucket, and was asked to write a dedication on a small piece of paper, 'From Eduardo Salazar, from the University of Edinburgh.'

I spent the days leading to the festivities just hanging around, having short conversations with people and getting to know the place. It quickly became obvious that Diamante suffered an extreme lack of public services. Most people scrambled home to lay out buckets and oil drums whenever it started raining and a queue of people holding towels and soap bars was a common sight at the main port entrance at sundown. The only source of relatively clean water for the whole community was a spring that had been put through a plastic tube near that entrance (see Fig. 5.2). The small trickle was everyone's shower and the whole community's water supply. I joined the queue on my second day and had to laugh at myself while showering with a tiny pink jug, as everyone found my incompetence in that set up quite entertaining.

Adding irony to this dire situation, the SRG had built outhouses with latrines and washing facilities for every household in the community, but they were mostly useless. As Duncan, a young community council member, told me, 'The taps and showers in them don't work, just the latrines. They've told us they would start working on water provision in June.' He had been nice enough to welcome me at his family's house where we talked about how *mitayo*, the hunting rounds some Yine men went on, were becoming increasingly harder as bushmeat became scarcer. We talked some more about his hopes with the road and the issue with the water, 'That road will be beneficial for us. Once we have it, the government will be better connected to our community and these improvements will happen quickly. We wouldn't be suffering because we don't have water.'

Fig. 5.2 The water supply in Diamante

Precarious water provision also seemed to be an issue for the nurse technician in charge of the health post. She was the only person that seemed to have time for a formal interview during Mother's Day preparations, the community council was happy to allow it. 'The water they have here is really bad. People need to get it with buckets out of the spring and that water can't be cleaned by just boiling it. It's worse during the dry season because the spring dries up, not a trickle comes out of the tube, and they have to use water from the big river. That's why the most common illnesses here are diarrhoea and stomach infections. They always come with that. We, the health staff, end up with the same issues as our patients. Even I am giving up already! We can't be buying bottled water all the time! I honestly prefer Tayakome [a Matsiguenka community inside the Park], where I was before, because an NGO has given them filters to clean their water. The health centre there also has 24 hour energy with solar panels.'

Not having clean water was just one example. The lack of availability of basic medical treatments and the difficulties in evacuating people to places where these were available further aggravated the effect of isolation on Diamante's health conditions. Carmen, in her own style, put things quite bluntly, 'If you have a medical emergency here you can live if you're lucky or die quickly.' Her comments reminded me of what Abelardo Gonzáles, mayor of Fitzcarrald [the district where all the off-road communities were] mentioned, '[In on-road communities] they call an ambulance, it comes, and they take the patient. Here if we call an ambulance it waits for us 30 km upriver and it takes two hours for us to get there on a boat. And that's only if it is daytime and a boat with a driver is available. If it's night time our patient will die. [With the road] we would see our injured and sick people surviving that trip.'

The nurse technician gave me some specific examples of this situation. 'There's an elderly man from Diamante who broke his foot by falling into a trench last night in Boca Manu. He was drunk. He needs an operation on his foot quickly but they can't do that there, much less here. His wife is trying to get a boat that will take them to Puerto Maldonado but she doesn't have enough money for that. It's so expensive! She doesn't even have money for accommodation or food there! The fuel alone will cost her about 400 soles.[6]' That was a significant amount of money for an older lady in a place where cash was scarce. The nurse continued, 'There was also a lady with complications at childbirth a few weeks ago. She hadn't been coming to her check-ups and we didn't know how the baby was because we don't have ultrasound here. When she started labour things got complicated because her uterus was swollen. We were lucky to find a boat on its way to Boca Manu so the obstetrician there could treat her. That woman could have died.'

On the night of the same day I had talked to her, I witnessed another example of this, which proved to be the harshest. The room I was staying at was right next to Celia's shop, where she had the only two pay phones in town. I recognized the nurse's voice, 'It's an anaemic girl doctor. Her mother brought her to the centre half an hour ago.', she sounded desperate, 'Her eyes have turned white! I have tried several things but I don't know what else I can do for her now. I don't know what to do!'

[6] About USD$ 120.00 at the time.

This night-time emergency had proven too complicated for the nurse technician and the resources she had. The child's evacuation needed to wait until morning. The nurse was trying to keep her alive until then. I listened to her despaired conversation, aghast, staring at the roof of my room in the dark. My fieldwork first aid course from university seemed useless. Soon after I woke up, I learnt the girl had died over night. The funeral took place a few days later, during the community's Mother's Day festivity, which was appalling to me but seemed to be part of daily life for Diamantinos.

5.3.2 *'That's All They Help Us with, Nothing More.'—The Immediate Effects of Connectivity*

The detrimental effects caused by the road in Shintuya had started happening in the 1970s, when the Peruvian legal framework to recognize Indigenous territories and safeguard Amazonian cultures was in its infancy (Greene 2009; Larsen 2016; Shepard et al. 2010). 'Things might be different in Shipetiari', I thought. After all, the road had only reached this Matsiguenka community 2015, after decades of pro-Indigenous legislations, and years after the '*Baguazo*', and event that cost lives but positively revolutionized Indigenous politics in Peru[7] (Greene 2009; Larsen 2016; Morel 2014; Royo-Villanova 2017). Additionally, most people I spoke to mentioned this community as an example of the road's success, the results of which I was expecting to see there.

I arrived in Puerto Shipetiari, the port across the river from the Community's settlement, after a noteworthy ride on a pickup truck I had barely been able to climb on as it left Salvación. I was expecting to see much of what so many people mentioned regarding the positive effects of the road there, but things did not look promising. The port was a small collection of shacks made with blue tarp and wooden beams surrounded by parked trailers full of timber. The buildings worked as restaurants, shops and lodgings at the same time, most of them run by Andean women and underage Matsiguenka girls as their assistants. A *trochero*[8] was transferring timber blocks to a larger trailer, similar to the ones used by banana wholesalers. Stacks of timber and towers of beer bottle cases were lying around in various places. Shortly after my arrival, I found Abdiel, a man from Shipetiari who agreed to take me to

[7]In June 2009, an ongoing protest against mining concessions by the Awajún people of the Amazonas region was forcefully suppressed by government police forces, resulting in several deaths on both sides (Royo-Villanova 2017). Following the 'dog in the manger' principles, this was portrayed by Alan García's government as an aggression guided by Indigenous leaders and organizations (Royo-Villanova 2017). It was later revealed as a plot by the government to provoke a violent reaction from the Indigenous protesters (Royo-Villanova 2017). This result legitimized the claim of national Indigenous organizations for the inclusion of consultation procedures for extractive concessions (Larsen 2016; Morel 2014; Royo-Villanova 2017).

[8]Literally, 'dirt tracker'. One of the old all-terrain Unimog trucks inherited from the time when the military built the older stretches of the road.

the community on his boat. He was making his way back there after an unsuccessful attempt to sell some yuccas to the restaurants.

Once I arrived at the community, finding my way to the Chief's house, as I had been instructed, proved very difficult for my untrained urban mind. Unlike Shintuya, the layout of this settlement still followed Indigenous Amazonian tradition and was guided by the streams and small rivers that went across it. I finally found my way to Magno Abelardo's complex thanks to the assistance of some community members I had been lucky enough to meet previously. I had to wait for him until he returned from his farm and met several other community members as they passed by his house.

Having the Chief as my first interviewee was quite revealing. As was expected by off-road communities, the new road was surely making travel less expensive. However, the cost reduction facilitated by the road didn't mean that travel was affordable yet. As Magno explained, 'The road has helped us just for travelling, which is still expensive. There are no affordable rates yet. We would like to have a bus like the one that goes to Shintuya but we still have to pay full price for a pickup truck to take us.' Ryan Alemán, who hosted me in Shipetiari, gave me more details later on, 'It's three hours from here to Salvación in a pickup truck. That costs us 35 soles[9] each. If I go with my wife, it's 70 soles each way. That is still too expensive and spaces on the vehicles aren't always available. We still have to beg for people to let us in their cars. We would need a bus for that to improve.' I thought about the local bus, which had begun serving communities along the Manu Road up to Shintuya out of the driver's own initiative about a decade after the road surface had been improved in the 1990s.

The enhancements to the community's health services, which off-road communities desperately looked forward to, hadn't been realized in Shipetiari either. According to the Chief, 'We have had our health centre for five years already but the staff doesn't know much. They still depend on help from the health centre in Itahuanía, but they help us only a little bit. They are quite stingy with their supplies; I guess they don't have many anyway.' The situation remained similar as in many off-road communities. In the absence of better healthcare, the community still had to evacuate their sick to other communities. Magno was more at ease when it came to this, 'Until 2005 we had to go to Shintuya on a boat in case of an emergency. It has been easier as the road came closer, even more now that it's just across the river.' The road had delivered a solution for medical evacuations in Shipetiari; having closer road access had driven death further away. Nonetheless, I would later learn about a woman from Shintuya who had died from complications at childbirth despite road connectivity. The road had allowed for her quick evacuation to Salvación but her condition proved to be beyond the capacity of the provincial capital's health centre. She had died on her way to Cusco. These musings also made me think of the time when my friend Ramón Ortega's wife had almost died because of a snake bite. The health centre in Salvación didn't have adequate facilities to store antivenin in, and Gabriela had to be evacuated to Paucartambo, in the highlands of Cusco, with few minutes of life

[9]Approximately USD$ 10.50 at the time.

to spare. This made me wonder if the improved chances of evacuating someone to Shintuya or Salvación were of much use in some extreme but likely situations.

Other expected changes hadn't taken place either, and the government authorities in charge of providing them still felt distant. Magno expressed his frustration at the lack of attention their requests for electrical connectivity got, 'It's not like before. We need energy. Now we want to use the internet, we want our cell phones to work while we're in our own community. But we need to go to Salvación to do all of that. We keep wondering 'when will they approve this?' It seems they never will.' He had not gotten clear answers about this from any government official. He had been told the substation in Salvación was going to be re-built eventually and almost nothing about the prospects of having the power lines connected from the road across the AMD river. The latter was a feat that had taken several years for Palotoa Llactapampa, an Andean community further upriver, to achieve. Improvements to the town's infrastructure also seemed to be an unfulfilled expectation for Magno, 'We have talked to the Manu Sub-Regional Government about getting their help to improve our port. We also want to urbanize our settlement, have streets. But they keep saying "no"'. Furthermore, Ryan talked about the communal meeting room in Shipetiari being half built. They were not getting the materials they needed to finish it. Apparently, the difficulties they had to get the SRG's attention were significant, 'Maybe if we go and bother them all day they'll help us. We need to go in person and demand things from them but sometimes we just don't have the money to go to Salvación and spend enough time there', was how Magno explained his concern. He later provided an example of the extent of the help they did get, 'We recently had a flood. Everywhere here was flooded. We made a list of everyone affected, took it [to the SRG], and they helped us but only with 15 sheets of corrugated tin roofing. That's what they help us with, nothing more.' His expression showed utter disappointment towards the authorities that had promised so much.

The highly expected economic opportunities that the road was supposed to provide hadn't materialized either. The volume of banana sales had not been anywhere close to what proponents of the road had assured. The price the Matsiguenkas of Shipetiari were being offered for the few bananas they did sell was low because their clients faced the risk of getting stuck on the road with their perishable cargo. As head of the community, Magno was clearly disillusioned, 'It was expensive for us to take our bananas [on a boat] to Itahuanía to sell them. That's why we thought having the road across the river would help us. But now that the road is there it hasn't done much for banana sales.' On the other hand, as my quick look around Puerto Shipetiari had revealed, timber sales were on the rise. Ryan, my host, compared the banana business with the timber business, 'We are seldom benefitted from our banana crops. With timber, on the other hand, we are not paying expensive boat fees anymore because we can just take it across the river and they'll buy it. We save a lot in transportation compared to before. It is cheaper to sell it and we can make a profit, which is what we wanted.'

Daniel Pérez, from the Manu Agricultural Agency[10] (MAA), explained what he thought was the cause of this when I visited his office in Salvación. 'The road is not getting built for technical reasons, to make calculated improvements or to increase sales of legal products. The authorities are using it for their political gain, for votes from all the loggers. There's no solid development plan for agriculture or public services coming with that road. Without these additional efforts surrounding the building project, that road will only exist for illegal timber extraction.' He reinforced his argument by describing what previous extensions and improvements of the road surface had done. 'The clearest result in the past has been the increase in illegal and uncontrolled logging. That is the only thing that the road has brought before. Communities just gave away their forestry resources. Shintuya has no more valuable timber to sell, for example. Now they eke a living out through fish farming and bananas. The same thing will happen all the way to Boca Manu and Colorado if this is not well managed, which is what seems to be happening.' He continued by explaining what was already happening in Shipetiari with the newest stretch of the road. 'Now Shipetiari is having the same issue as Shintuya did; well-known loggers have gone there and offered the moon and the stars and community members have started selling their trees for cheap. Before the road got there, they had cedars and other hardwoods even within their town but now that's gone, they've sold it. And what did that do for them? I would say they're worse off than before. I'm afraid banana farming is going to be their only resource soon.' It seemed as if history would repeat itself in the newest community connected to the road, despite the legal or political changes that had taken place at a national level since the 1970s.

Not only that, this time, it seemed, the transformation would be even riskier. My former colleague Paolo, who worked for an NGO that supported Shipetiari directly, told me at a meeting we had in Cusco that Puerto Shipetiari and the road that continued from there were only used for illegal timber extraction and for fuel trafficking that served the illegal mines down the Madre de Dios river, and quite openly so. 'The authorities must know about it as it is so evident. But they haven't reacted yet. Or, worse even, they are making profit from it as well. The great winners with all of this are the loggers', he told me. Magno and Ryan knew about this illegal trade as well. 'We've had a meeting with Nuevo Edén once because they were transporting illegal cargo on the road and the port. We have some outsiders using the port, strangers, and because transport is through the river here there is no real safety of who comes and goes', said Magno. 'Unknown people come sometimes, they transport illegal fuel. That's a problem too. For a long while many trucks were bringing fuel. They call it 'ambulatory fuel', because it's not authorized for sale. They demand our help in their dealings and we have no way of recurring to the law to avoid that', was what Ryan said about it, 'We don't know what they do, they may be doing something illegal that might involve us later. There's been a lot of that sort of unknown characters in Puerto Shipetiari. They go further down river, they build their small settlements and work on things they shouldn't be working on. We're scared of that. We can end up being involved in that as a community. We try to be very careful.' Ryan's premonition came

[10]*Agencia Agraria Manu.*

true a year after our interview, when a small airplane carrying almost 300 kilos of cocaine was seized at a clandestine airstrip within the community's territory (Exitosa Noticias 2018).

5.4 Road Connectivity and Isolation—Unpacking the Promise of Political Integration

The following two subsections evaluate the extent to which the 'enchantment' of political integration is present in off-road communities in Manu (Harvey and Knox 2012). As with the sections focused on the 'enchantment' of speed, here we analyse the aspirations held in off-road communities in Manu with the realizations of people in on-road communities. Again, the mismatch between aspirations and realizations regarding the road manifests the presence of the 'enchantment' of political integration.

The expansion of the state's authority to isolated and thinly populated areas of the national territory has been the primordial motivation of efforts like the government-mandated 'colonization' of the Amazon in the 1960's in Peru and elsewhere (Campbell 2012; Greene 2009; Harvey and Knox 2015; Morel 2014). The most cost-effective tool to exercise land sovereignty in areas like these is road connectivity (Barrantes et al. 2014). Apart from territorial control, the speed provided by roads is expected to facilitate access for government officials in charge of addressing issues that concern their constituents, as well as allowing safer and more efficient access for public service providers (Campbell 2010; Harvey and Knox 2012). Improving the quality of life in communities through the provision of electricity, water, and other public services is also assumed to be more feasible with easier access (Harvey and Knox 2012, 2015). The remoteness of frontier areas like the Madre de Dios basin and the conditions that communities in them face instil feelings of abandonment and political isolation, especially in off-road communities (Harvey and Knox 2008, 2012, 2015; Murray-Li 2014; Uribe 2017; Wilson 2004). As a result, the expectations that these communities have of the state once roads are built are quite high (Harvey and Knox 2008, 2012, 2015; Murray-Li 2014; Uribe 2017; Wilson 2004). However, as with the failure to fulfil the promise of speed, and also because of it, the promised political integration is scarcely delivered in most cases.

5.4.1 A Sense of Belonging—Aspirations of Political Integration

One of the main justifications for the 2013 Law Proposal was the increased access that the local authorities of Manu would have to their regional capital, Puerto Maldonado, through the town of Boca Colorado (see Map 2.1) (Congreso de la República 2013).

An almost two-day trip through the city of Cusco and the Interoceanic Highway is currently required for any paperwork or personal consultations with GOREMAD, something that private citizens also face when they require official documents from the regional authority (Congreso de la República 2013). The alternative is a similarly long but much riskier and more expensive boat trip along the Alto and Bajo Madre de Dios rivers. Increased access for authorities in Salvación would work the other way around as well. When speaking to SRG officials, the expectation that the road would make it easier and cheaper for them to visit previously off-road communities and improve state-provided services in them was frequently put forward.

The utter frustration that off-road communities feel about the lack of government presence is seen clearly in the sentiments held by the people of the neighbouring communities of Isla de los Valles and Boca Manu. As Augusto Ramirez, chief of Isla de los Valles, put it, 'Our community has always been forgotten by our authorities. [Their support] is next to nothing, they don't come here because [boat] transport is so expensive.' Testimonies like this one also show how the Manu Road symbolizes the presence of the state and a sense of belonging to the country that gains so much from the touristic resources of nearby protected areas. In off-road communities, which only have two hours of diesel-generator power each day, the road was also seen as the entry way for energy provision. This was especially so because of the recent extension of the power lines, mentioned in Sect. 5.1.3, which foreboded the arrival of electricity. The critical situation that Diamante faced, described in Sect. 5.3.1, is also evidence of this illusory element. A community of three hundred people that depended on a single trickle of unpurified water and whose houses all had been given inoperative washing facilities, shows how desperate circumstances motivate the faith in the arrival of government authorities through the road. The water quality and availability in Boca Manu and Isla de los Valles were not much better either.

The expectations were similar for the case of public service providers. As mentioned earlier, the nurse in Diamante was ready to quit her job because of the appalling conditions she worked in. As Carmen, the community's Vice-chief, told Eduardo, 'Health professionals don't want to come to my community because of how difficult transport is.' In Boca Manu, *Profesor* Paniagua told Eduardo about the dire consequences teachers in Boca Manu worked under, 'We have to travel risking our lives to get here. The little money we make we can only cash in Salvación and travelling there is too expensive. The cost reduces what we have left to feed ourselves and our families back home with. Many teachers have been in fatal accidents as well, not just on boats but on the road. It is dangerous! Our life is practical martyrdom!' Workers from the local boarding house for Indigenous students also mentioned that, because of this situation, two teachers had quit before the year's first month of classes ended. The school in Boca Manu was left only with untrained staff who had no other choice but to work in these challenging conditions. The fact that remoteness generates teacher absenteeism in this area is well recognized (INRENA 2008). For Paniagua, Carmen and many others, difficult access to off-road communities resulted in understaffed government health centres and schools. Thus, the road was also expected to make it easier for government teachers and medical staff to provide their services, which made *Profesor* Paniagua a staunch supporter of the road.

Furthermore, it was believed that the road would also make it easier for people from these communities to access government services and benefits elsewhere. For Duncan from Diamante, the road would make it easier for young people to access higher education, 'The road would allow young people to continue studying. They could get out of here and study in Cusco, have new ideas and bring them back here.' For Augusto from Isla de los Valles, the road would make it safer for the elderly to cash their pension checks in Salvación. Others also mentioned the possibility of going to Salvación to liaise with authorities, a benefit of roads that is frequently considered (Harvey and Knox 2012).

Additionally, government officials added law enforcement to the equation: by facilitating access, the Manu Road would make it easier for government agencies to enforce the law in previously off-road communities. Moreover, this was occasionally mentioned with regard to the enforcement of environmental law, an idea that is recently being used to justify road construction in the Brazilian Amazon (Campbell 2012). According to these authorities, the lack of a road made law enforcement impractical, even impossible, in off-road communities.

The construction of the road, then, appeared as a requirement and a means to access services that off-road communities had not had before, like water provision and electricity, as well as improving existing services like education and healthcare. The road would also facilitate the enforcement of environmental law and improve the possibilities for those who wanted to access government services and benefits outside their communities. The abandonment felt by off-road communities was expected to end with road connectivity.

5.4.2 Unfruitful Political Integration

As seen in the accounts presented previously, even when Shintuya had been connected to the road network for decades and Shipetiari had had two years of connectivity, state-provided services were still lacking and the political weakness common in the frontier seemed to remain (Sunderlin et al. 2005). The 'friction' that alters the expected effects of infrastructure in the frontier was evident in these communities (Harvey and Knox 2012; Tsing 2005).

When asking different community members in Shintuya about what positive changes had taken place there since the construction of the road, most did not give a positive outlook. The slow and meagre nature of the effects that political integration had produced was evident. The changes that modernity had brought to Shintuya had been implemented much after the arrival of the road, even after incomers had come and gone, and had made the most out of their capacity to appropriate the benefits the road could have generated for locals (Harvey et al. 2017; Larsen 2016; Ráez Luna 2017). Furthermore, Shintuyans were not satisfied with the perceived imposition of some of the 'improvements' they had been given, like their modern outhouses and sewage system, which seemed to disturb the few cultural practices that remained. Besides, many of these changes had depended on the community's benefactors: the

Dominican mission and Hunt Oil, not on assistance by the state. Would off-road communities require the presence and intervention of similar actors to accomplish their aspirations once connected to the road? What would they need to give in exchange? On the other hand, some of the advancements that the state had indeed provided were implemented piecemeal. The most recent example of this was the inoperative power lines that went through the community, which seemed to stand as monuments to its remaining political disintegration. Leaving Shintuya through the road was not a significantly beneficial option either. The few young people whose families had the money to send them elsewhere faced great economic costs while living in the cities, which prevented them from concluding their studies. Those who left to seek work were subject to the abuses of labouring for illegal activities, like mining and logging, or bared the brunt of being included as the lowest piece of the urban job market. The unpredictable and illusory outcomes of connecting the frontier with the global meant that aspirations like those held by young Duncan, that youth from Diamante would educate themselves in the cities and return to their communities to improve them, were seldom realized (Harvey and Knox 2012; Tsing 2005).

Magno Abelardo spoke about a similar situation regarding the fate of the youth of Shipetiari, 'We count on our young students to keep defending our native community but sometimes students don't come back from studying because they have to work in the cities. After that they don't think of coming back to their community.' As seen in Sect. 5.3.2., the conditions in Shipetiari after two years of road connectivity were not promising either. For example, the improvements to health services that off-road communities expected were clearly absent. The local health centre was as understaffed and undersupplied as Diamante's, perpetuating health risks and the dependence on services provided in other communities. The danger that this Matsiguenka community had faced from violent raids by isolated groups of nomadic Yine,[11] which had already caused the death of a young man shot by an arrow, had apparently not been enough motivation for the government to improve emergency healthcare capabilities (García Delgado 2018). And even if more staff and equipment were available, the balance of the situation might not have been positive as, according to the head of healthcare of Manu province, the road was already facilitating the transmission of locally rare illnesses that most local health centres struggled to treat. Other services available had not been a benefit of the road and weren't supplemented by it, as Shipetiari had had running water for some time before the road arrived (INRENA 2008), but the toilet facilities in most households remained as open-air latrines by the time of Eduardo's visit. Furthermore, the hopes that Magno Abelardo had for energy provision and mobile phone connectivity for his community seemed only to frustrate him. The uselessness of the power lines along the road added to his worries about the need to build towers to put those lines through the AMD river to

[11]The 'Mashco-Piro', literally 'wild' Piros, are a sub-group of the Yine ethnicity, also called Piro, that separated from its core after the violence generated by the rubber trade in the early twentieth century. Small groups of people from this 'voluntarily isolated' ethnic sub-group have been establishing contact with settled Indigenous communities, preachers and government officials in the last decade, which has had fatal consequences in a few occasions (AIDESEP 2018; García Delgado 2018; Torres López 2016).

reach their settlement. Nonetheless, what reflected the lack of political connectivity that Shipetiari still suffered the most was that lack of attention their pleas for assistance with communal infrastructure got. Their half-built communal meeting room and the 15 sheets of roofing they had received as post-flood aid said much about this situation. The fruitless and expensive efforts to 'bother' the authorities for their attention indicated that the issue was not one of physical integration but of an actual lack of interest from the authorities in the wellbeing of these communities. The road still did not allow community leaders to effectively make their government accountable, as is often supposed to happen, which kept their communities politically weak (Harvey and Knox 2012; Sunderlin et al. 2005).

Another example of the ineffectiveness of political integration was the lack of law enforcement across the river from Shipetiari. The amount of timber Eduardo saw in Puerto Shipetiari, and the openness with which logging operations were conducted, also revealed a lack of political will to enforce the laws that affected the logging trade in the province. This was clarified by what was mentioned by Dámaso, owner of a timber mill in Salvación, 'Very few people work within the limits of their logging permits. Timber here is illegal, 'pirated'. They take it out in whatever way they can.' This is exactly what happens with most forestry operations throughout the country (Global Witness 2019; Urrunaga et al. 2012, 2018). Despite this nation-wide condition, the absence of the Peruvian Forestry Service in the AMD river basin and SERNANP's lack of resources facilitated this situation. The clear absence of law enforcement despite road connectivity exemplifies the capacity of frontier roads to foster illegal activities (Harvey and Knox 2015; Tsing 2005). Moreover, Salvación, as the administrative capital of the province, also fostered a widely recognized level of corruption in the local forestry office, in charge of controlling the logging permits mentioned by Dámaso. Eduardo witnessed several examples of this. He recounts:

> In my previous experience in Manu, I had heard the stories of 'Santa Rosita', a very permissive former forestry official nicknamed after the patron saint of Lima who appeared in the 200 sol[12] notes. This time, after visiting the MAA, I saw the current forestry official having a very animated conversation with two drivers standing in between two lorries full of timber. The conversation stopped while I passed by.

These examples of corruption show the contrast between the demand for political integration and the resistance against the enforcement of law seen in the frontier in similar cases (Sunderlin et al. 2005; Wilson 2004). They are another reflection of the widespread corruption that affects the Peruvian forestry sector as well (Global Witness 2019; Urrunaga et al. 2012, 2018). The current situation in on-road areas made it clear that a key issue for the enforcement of environmental law was not lack of access but lack of resources, political will and corruption.

As with the promise of speed, and partly because of its lack of fulfilment, on-road communities were still not benefitted by political integration and remained weak in this regard (Harvey and Knox 2012; Sunderlin et al. 2005). This was worsened by an apparent lack of political interest in the wellbeing of these on-road Indigenous communities, and by a lack of resources and political will to enforce the laws that

[12]Approximately USD$ 60.00 at the time.

would affect the local timber trade. However, despite witnessing this situation, off-road communities still trusted that the government would bring dramatic changes to their lives through the road, which was a manifestation of the 'enchantment' of political integration.

Overall, we have examined how the 'enchantments' of speed and political integration take place in the communities in Manu, and the mismatch between off-road aspirations and on-road realizations (Harvey and Knox 2012). The aspirations for increased speed in travel and improved integration with political centres are certainly key in the imaginary of the road held in off-road communities. However, as we will see in the following chapter, the aspirations of economic connectivity seem to be the strongest and, consequentially, the power of the road to 'enchant' in this respect is the most distinct.

References

AIDESEP (2018) Informe sobre la situación de los pueblos en aislamiento y contacto inicial de la Amazonía peruana. Lima, AIDESEP

Barbier EB (2012) Scarcity, frontiers and development. Geogr J 178(2):110–122

Barrantes R, Fiestas J, Hopkins Á (2014) Evolución de la infraestructura de Transporte y Energía en la Amazonía Peruana. In: Barrantes R, Glave M (eds) Amazonía Peruana y Desarrollo Económico. GRADE, IEP, Lima, pp 109–160

Barrantes R, Glave M (2014) Introducción. In: Barrantes R, Glave M (eds) Amazonía Peruana y Desarrollo Económico. GRADE, IEP, Lima, pp 13–20

Campbell B (2010) Rhetorical routes for development: A road project in Nepal. Contemp South Asia 18(3):267–279

Campbell JM (2012) Between the material and the figural road: the incompleteness of colonial geographies in Amazonia. Mobilities 7(4):481–500

Congreso de la República (2013) Proyecto de ley que Declara de Necesidad Pública e Interés Nacional la Construcción de la Carretera de Integración Regional de Madre de Dios. Comisión de Transportes y Comunicaciones, Lima

Exitosa Noticias (2018) Madre de Dios: Policía interviene avioneta con más de 293 kilos de cocaína. https://exitosanoticias.pe/policia-interviene-avioneta/

Figallo M, Vergara K (2014) La Amazonía Peruana de Hoy. In: Barrantes R, Glave M (eds) Amazonía Peruana y Desarrollo Económico. GRADE, IEP, Lima, pp 47–108

García DF (2018) Comunero muere por impacto de flechas de indígenas mashco piro. El Comercio Website. https://elcomercio.pe/peru/madre-de-dios/contactados-matan-flechazos-comunero-peru-noticia-501491

Global Witness (2019) El justiciero forestal: Por qué se debe devolver la independencia a OSINFOR y expandir sus poderes. Global Witness, London

Greene S (2009) Caminos y Carretera: Acostumbrando la indigenidad en la selva peruana. IEP Instituto de Estudios Peruanos, Lima

Harvey P, Jensen CB, Morita A (2017) Introduction: infrastructural complications. Infrastructures and social complexity: a companion. Routledge, London, pp 1–42

Harvey P, Knox H (2008) Otherwise Engaged. J Cul Econ 1(1):79–92

Harvey P, Knox H (2012) The enchantments of infrastructure. Mobilities 7(4):521–536

Harvey P, Knox H (2015) Roads: an anthropology of infrastructure and expertise. Expertise, Ithaca, N.Y

INRENA (2008) Plan Maestro de la Reserva Comunal Amarakaeri 2008–2012. Ministerio de Agricultura, Lima

Kernaghan R (2012) Furrows and walls, or the legal topography of a frontier road in Peru. Mobilities 7(4):501–520

Larrea-Gallegos G, Vazquez-Rowe I, Gallice G (2016) Life cycle assessment of the construction of an unpaved road in an undisturbed tropical rainforest area in the vicinity of Manu National Park, Peru. Int J Life Cycle Assess 11(17/2016):1–16

Larsen PB (2016) Derechos indígenas, gobernanza ambiental y recursos en la Amazonía peruana, Hacia una antropología de la posfrontera. Instituto de Estudios Peruanos, Lima

Leal DB, Salisbury DS, Silva J (2015) Ideas cambiantes sobre territorio, recursos y redes politicas en la Amazonía indigena: un estudio de caso sobre Perú. J Lat Am Geogr 14(2):181–204

Morel J (2014) De una a muchas Amazonías: Los Discursos sobre 'La Selva' (1963–2012). In: Barrantes R, Glave M (eds) Amazonía Peruana y Desarrollo Económico. GRADE, IEP, Lima, pp 21–46

Murray-Li T (2014) Land's end: capitalist relations on an indigenous frontier. Duke University Press, Durham and London

Oliart P, Biffi V (2010) Territorialidad indígena, conservación y desarrollo: Discursos sobre la biodiversidad en la Amazonía peruana. Instituto del Bien Común, Lima

Pieck SK (2011) Beyond post development: civic responses to regional integration in the Amazon. J Cult Geogr 28(1):179–202

Pieck SK (2013) Asphalt dreams: road construction and environmental citizenship in Peru. Dev Chang 44(5):1039–1063

Ráez Luna E (2017) Pueblos Indígenas y Conservación de la Naturaleza: ¿Debates teóricos o urgencias vitales? El caso del Parque Nacional del Manu. In: Ráez Luna E (ed) Derechos Ancestrales y Conservación de la Naturaleza en Debate: El caso de los Pueblos Indígenas del Parque Nacional del Manu. Seminario Permanente de Investigación Agraria (SEPIA), Lima

Royo-Villanova J (2017) La otra cara del baguazo. Planeta, Lima

Salisbury DS, López JB, Vela Alvarado JW (2011) Transboundary political ecology in Amazonia: History, culture, and conflicts of the borderland Asháninka. J Cul Geogr 28(1):147–177

Shepard GH, Rummenhoeller K, Ohl-Schacherer J, Yu DW (2010) Trouble in Paradise: indigenous populations, anthropological policies, and biodiversity conservation in manu national park Peru. J Sustain For 29(2–4):252–301

SPDA (2017) Reportan pérdida de 1660 hectáreas de bosques amazónicos debido a vientos huracanados. http://www.actualidadambiental.pe/?p=43447

Sunderlin WD, Belcher B, Santoso L, Angelsen A, Burgers P, Nasi R, Wunder S (2005) Livelihoods, forests, and conservation in developing countries: An overview. World Development 33(9 SPEC ISS):1383–1402

Torres LF (2016) The highway tearing through the heart of the Amazon. Ojo Público, 1–22. http://ojo-publico.com/sites/apps/amarakaeri-parte1-english/

Tsing AL (2005) Friction: an ethnography of global connection. Princeton University Press, New Jersey

Tsing AL (2009) How to make resources in order to destroy them (and then save them?) on the salvage frontier. In Histories of the future (pp 51–73)

Uribe S (2017) Frontier road: power, history and the everyday state in the colombian amazon. Wiley, New Jersey

Urrunaga J, Johnson A, Orbegoso I, Mulligan F (2012) La máquina lavadora: Cómo el Fraude y la Corrupción en el sistema de concesiones están Destruyendo el Futuro de los bosques del Perú. Environmental Investigation Agency, London, Washington DC

Urrunaga J, Johnson A, Orbegozo D (2018) El Momento de la Verdad: Oportunidad o Amenaza para la Amazonía Peruana en la lucha contra el Comercio de la Madera Ilegal. Environmental Investigation Agency, London, Washington DC

Vadjunec JM, Schmink M, Greiner AL, Vadjunec JM, Schmink M, New ALG (2011) New Amazonian geographies: emerging identities and landscapes. J Cul Geogr 3631(1):1–20
Wilson F (2004) Towards a political economy of roads: Experiences from Peru. Dev Chang 35(3):525–546

Chapter 6
The 'Enchantment' of Economic Connectivity

Abstract After exploring the way in which the 'enchantments' of speed and political integration manifest in off-road communities in Manu, we now turn to analyse if and how the 'enchantment' of economic connectivity is present in this case. Economic connectivity, the possibility of increasing local economic growth through trade relations boosted by road access, is the most highly expected and frequently mentioned 'promise' of the road, both in local communities and the development discourse. In the following sections Eduardo's fieldwork accounts reveal key details regarding the fulfilment of this aspiration for economic growth. Afterwards, we analyse the aspirations of off-road communities and the realizations of communities with road access regarding this element. It seems, once more, that the possibility of fulfilling the promise of economic connectivity based on activities not related to timber is slight at best.

Keywords Manu road · First roads · Frontier infrastructure · Enchantments · Economic connectivity · Logging · Political ecology

6.1 Stories of the 'Enchantment' of Economic Connectivity

The following accounts describe some of the situations that Eduardo encountered which illustrate the way in which the 'enchantment' of economic connectivity manifests in Manu. In them, the strong aspirations for an increase in the banana trade in off-road communities are clear, but the complexities of achieving this are also evident.

6.1.1 'One Raceme is Nothing'—The Economic Costs of Isolation

Apart from illegal timber extraction, banana farming is Manu's single most prevalent economic activity (INRENA 2008). During my two years working for an NGO in the area, I didn't see any other produce being farmed for commercial purposes. Local

E. Salazar Moreira and M. Palomino-Schalscha, *Road Expansion in the Peruvian Amazon*, SpringerBriefs in Latin American Studies, https://doi.org/10.1007/978-3-030-47182-8_6

farmers had various stories that explained this. Sugar cane plantations and cane-liquor production seemed to have come and gone with the large estates that were confiscated by the Peruvian government after agrarian reform in the late 1960s, it was the stuff of legend. Farmers' attempts to grow cacao had failed in decades past because of a plague that had become infamous, farmers tended to stay away from it. Coffee had suffered a similar fate. Papayas and pineapples would also be destroyed by plagues if a large amount of pesticides was not used, which was unaffordable for most. Yuccas were too fragile to make the rough day-long trip to Cusco. Some of these products were farmed for subsistence at a very small scale, and abandoned patches of them, remnants of old plantations, were a welcome surprise in people's forays into the degraded forest.

The banana was king of the farm in on-road communities, plantains were a staple food for locals. Despite the occasional threat of banana-specific plagues, the MAA didn't incentivize diversification and kept focusing its meagre efforts on this crop. The variety of competitions at the yearly farming fair had diminished with time and the only prize that was still given went to the heaviest banana raceme. The queen of the local beauty pageant was 'Miss *Plátano*', a fiercely contested title among young women from Salvación to Boca Manu. The smell of the rotting bananas left by wholesalers on their way to Cusco had become a favourite of mine, an acquired taste.

Despite the lack of crop diversity, the banana trade allowed families in on-road communities to make a profit and invest in their wellbeing. Some, of course, were also involved in logging to a certain extent but farming allowed them to have a more reliable, and legal, source of income. Many invested in small vehicles to bring their harvest to the road, others bought land, and some worked with various NGOs to try out new 'sustainable' farming practices. A few who missed their days as loggers even began to delve into agroforestry, dreaming of chopping down big trees in the decades to come and receiving large amounts of money for their timber in one go. Furthermore, an achievement that those more successful could boast about was being able to send their children to study in the cities. They would become professionals, technicians at least. Many dreamed of people calling their sons 'Mr. Engineer', a title only the highest ranked SGR officials had and were recognized for. For those who did not have a profession or could not run a logging operation, banana farming was a way to make a stable living.

In contrast, things were not the same for farmers in off-road communities. In my time in Isla de los Valles, Jesús Ríos, vice-chief of this Yine community, sounded truly upset when telling me about their situation. 'We have products, we are farmers, but we have no way to sell a papaya, a banana, a pumpkin. We can't sell products to have the money we need to give our children when they go to school, they don't have notebooks or pencils'. After mentioning their own ordeals, the lack of school supplies that the students faced was frequently mentioned by teachers at the Boca Manu school, where kids from Isla de los Valles also attended. Their nourishment with commercial foodstuffs seemed to be another concern. Jesús continued, 'We suffer to buy many things to eat, like sugar, rice. We don't have any way to sell our fruit! Good harvests rot as we are the only ones who eat from them.' For him, the road was the way out of this confinement, 'Other communities like Shipetiari, that are connected to the road, can sell their products. They are at ease, they have enough to keep their children in school. On the other hand, we live in extreme poverty! Boat travel is too expensive to take our fruit elsewhere, and we don't have boats or motors of our own. How are we going to sell our products? It's like being in a cage! We are stuck here!'

Others in neighbouring Boca Manu had a similar outlook and also mentioned on-road communities as examples of success. 'Communities that have road access have trucks arriving constantly to buy their timber, their yuccas and bananas. They have economic movement. We are expecting to have an income. That is the vision that we have for when the road reaches us', said Luis Manuel when I interviewed him at his restaurant in Boca Manu. 'I'm also expecting to have movement in my establishment when that happens', the place was notoriously empty when I spoke to him there. It seemed overly large even when I tried to imagine everyone in town inside it.

Duncan, from Diamante, shared the same point of view. 'The greatest benefit the road will give us is that the community will sell their products. We will be able to work our land more and make money from it. Everyone here grows bananas, plantains, yuccas, papayas. We would like to sell those products. Sometimes the tourists pass by and buy a raceme, but that's all. One raceme is nothing. That's why the community has already decided that it wants the road'. The idea that the road would improve the economic prospects of off-road communities seemed to be shared unanimously by locals.

6.1.2 'We Risk Our Cargo and Our Lives to Come!'—An Unviable Banana Market

In order to see if the dream of making a living from agriculture was realistic, I had to consider what the current banana market in Manu looked like. There was nowhere better along the Manu Road to inquire about potential banana sales than Santa Cruz, the port settlement where farmers from Palotoa-Llactapampa sold their bananas. This community of Andean migrants, all devout Adventists, was recognized by everyone in Manu as an example of hard and well-organized agricultural work.

After a two-hour ride from Salvación on Mr. Cabello's bus, I arrived in Santa Cruz and walked down the side road to the port entrance. As I walked across the stony riverbed towards the banana trailers, I saw the most active river crossing movement I had ever seen, even after working for years in Manu. Longtail shaft boats, nick-named *peque-peques* throughout Peru, were coming from the Palotoa side so full of racemes that they seemed one banana away from sinking. People started unloading the racemes even before the boats stopped, they turned to drive back as soon as the last one was lifted from the hull. An old associate of mine, Mr. Egidio Pacsi, greeted me. He presided over the Palotoa Producer's Association and had recently been appointed as president of the Manu Agricultural Federation, something which he proudly announced. He was unloading racemes from a boat himself and grouping them according to their quality for washing and appraisal (see Fig. 6.1). The banana business was like clockwork in Palotoa. Egidio looked proud of his work and that of his fellow farmers, he knew they were an example of the economic potential of agriculture in the Manu province.

My familiarity with him eased my conversation with the wholesalers. 'This road is dangerous enough from Paucartambo to here', said Gumercinda, the Andean woman in charge of the first truck I approached. She showed me how the tarp that covered her cargo hold had been torn almost fully by a falling bamboo stem on the way down through the cloud forest. The danger of having a sharp bamboo shaft falling on her truck while driving on a narrow and muddy road on the side of the almost vertically steep mountainside was quite clear to me, 'We risk our cargo and our lives to come!' After my own near-death experiences on the road from Cusco, I had no doubt about that.

When I inquired about the possibility of driving further to buy bananas from other communities, Gumercinda explained her risk-assessment process. She considered Palotoans to be better organized and more trustworthy, 'Down the road the bananas aren't even harvested for you, here they wait for us with everything ready. It wouldn't be worth the trip otherwise'. She continued explaining how she would not even consider going further until the road to Boca Manu was tarmacked, bridges were built, and they were certain of the quality of the bananas being sold. 'There's no quality further down, they just sell second-class bananas. They rarely come out with big fat ones, most of them are skinny and the racemes are shorter', she added.

Fig. 6.1 Palotoans unload banana racemes in Santa Cruz

All the other wholesalers whom I spoke to agreed that going further, to Shipetiari, even to Shintuya, was considered too risky because the condition of the road worsened. 'The road is abandoned. The authorities here don't think of what people need. The Serjali and the Mochina [rivers, between Shintuya and Itahuanía] fill up quickly and that makes the trip further down too risky', was the way one of them put it. Like Gumercinda, most drivers considered that off-road communities would still be too far away to be worth travelling to when the road was built. And even if they did get to these communities, the price most of them would be willing to pay per raceme would have been far lower, considering that it dropped from 15 soles[1] in Salvación to 7 soles in Shipetiari. After some time with them, I had to run back up to the road. Mr. Cabello was honking his horn, on his way back to Salvación already. He was famous for not waiting for too long, so I only had the chance to wave Egidio goodbye.

In Salvación I also spoke to Ricky and Nidia, a young couple who buy bananas in Shintuya. They were the first to attempt to establish a firm relationship with banana producers from this community around 2010. I chased after them for quite a long time. Their house seemed to be empty all the time because of their frequent travel in and out of Manu. I tried to make them aware of my interest in talking to them in many ways until I saw their truck parked outside their house one evening. They welcomed me saying they were expecting me; my interest had become small-town gossip. I was lucky, they were leaving on another round early the following day.

They spoke of the long process they had gone through to establish themselves as wholesalers in Shintuya, which was not something many others were up for. They started out buying bananas from Andean migrant communities close to Salvación, like Águanos and Mansilla, but soon found that other wholesalers had already taken over the production there. Farmers had formed trustful relations with earlier buyers, which was a barrier for their entry. The situation was worse for them further down the road in Santa Cruz. So, they had the idea to buy from Shintuya and boost the market with the Harakmbut people there. Wholesalers who had arrived before them paid too little and were not consistent with their visits. 'We started out with a higher price, so people began to awaken. Others would go and try to take advantage of them. They would pay 7 or 9 soles[2] for each raceme. We offered them 12 or even 14 soles,[3] which is what was offered in communities closer to Salvación', was the way Nidia explained it.

She was the daughter of a former work colleague of mine, Mr. Ramón Ortega, who had been experimenting with sustainable farming practices in Salvación for several years and was respected in Manu as one of the few Andean migrants who had always respected Indigenous farming practices. He had also been a schoolteacher in some of the communities in the area. 'They recognized us because of my father's work. The community accepted us and we have worked with them ever since. That's why they call my husband "Papá Ricky"'. Their effort at generating a market for Shintuya's bananas had helped them earn enough to buy a larger and newer truck,

[1] About USD$ 4.50 at the time.

[2] Around USD$ 2.40 at the time.

[3] Around USD$ 3.90 at the time.

which Ricky drove proudly. He continued explaining their efforts, 'Most drivers won't buy bananas that are starting to ripen but we do. We buy all the bananas we're offered and then give away or throw away the ones we think won't make it to Cusco. We do that so they won't feel bad. Their tradition makes them very sensitive to rejection so it's better to buy what they offer. Otherwise they won't want to sell to you because "we're not friends anymore". You can't reject what they offer'.

Considering their willingness to make an extra effort to buy from Indigenous people, I asked if they would consider driving the extra 25 km to Diamante or little over 30 km to Boca Manu when the road was built. 'No, it's too far, there's enough production much closer. We wouldn't consider that unless the quality decreased closer by, which happens sometimes when the soil isn't good anymore. We'd have to look for better bananas or no one in Cusco would buy them. That would be dangerous for us because we can't risk losing even one round, that would compromise our capital. We need to look out for that, it's not just buying fruit'. Ricky's explanation of the financial risk banana wholesalers faced when losing a load clarified the reluctance to drive further that most of the ones I had met in Santa Cruz shared. Nidia added that the only reasonable incentive to drive further down the Manu Road would have been timber, 'If you're a logger and your trailer gets stuck on the way, the only thing you need to do is unload and leave your cargo on the side of the road. Timber can be left out in the open for months and won't get damaged, it will not rot like our bananas do in just one day'.

How did people in off-road communities figure wholesalers would buy bananas from them? All of them used the road at one point or another and had contact with people from on-road communities. Certainly, they had talked about the prospects of the banana market and other expectations regarding the road.

6.2 A Market for Bananas, or for Timber?—Examining the 'Enchantment' of Economic Connectivity

The expansion of global markets in the last several decades has given road connectivity in the frontier a new role apart from political integration: connecting communities to larger markets (Campbell 2012; Carvalho et al. 2002; Pieck 2011). Lack of road access is seen as a barrier to development because it prevents integration to the global market and the economic growth that is expected from that (Jarrier 2012; Pieck 2011; Quinhoes 2011). This is the main justification for current road expansion policies and transport infrastructure projects, like IIRSA and the smaller undertakings that stem from it (Barrantes et al. 2014; Jarrier 2012; Quinhoes 2011). Given the rhetoric that promotes these modernizing projects, the most anticipated promise that road building and improvement make to frontier communities is that of increased economic connectivity (Harvey and Knox 2012). The aspiration held by these communities is that roads will allow profitable trade by providing locals with better access to formerly inaccessible markets, and also by giving outsiders access to resources previously out of their reach. As seen in the testimonies gathered by Eduardo, the main aspiration that off-road communities in Manu hold is to be able to sell the bananas they grow, to welcome more tourists, and to pay lower prices for the consumer goods they need. The road is also expected to generate a cost reduction for the timber business but this is not often mentioned, or is even denied. Nonetheless, as with previous promises of the road, the lack of enough speed prevents most expected changes from taking place and seems to mostly benefit logging, an already profitable and illegal business. We now move on to explore the aspirations that off-road communities in Manu have regarding economic connectivity with road access.

6.2.1 The Trailers will Come—Aspirations of Economic Connectivity

The aspirations of economic connectivity generated by the Manu Road are evident in the comments made by informants in Isla de los Valles, Diamante and Boca Manu, described in the stories narrated earlier. Inspired by their overwhelming aspirations and what they saw in some on-road communities, they envisioned trailers arriving to buy their agricultural produce and hardwoods, motorcycle taxis bringing curious and adventurous tourists, as well as cheaper food and school supplies at the local shop. This is also expressed in the 2013 Law Proposal for this road, which places better integration to the agricultural and tourism markets as the main benefits that would be generated by this infrastructure (Congreso de la República 2013). The road would increase local incomes by opening access to previously scarce economic opportunities through the reduction of transportation costs. Informants in off-road communities thought the lack of access to such opportunities was the source of their poverty and mentioned Shipetiari as an example of the expected commercial success facilitated by the road.

The product most talked about was bananas, a crop all communities in the area grow for subsistence and the predominant agricultural produce of Manu's on-road communities (INRENA 2008). As Duncan from Diamante told me, the current demand for them was almost nonexistent in those off-road. His faith in the banana business was based on the belief that agriculture was something all community members could do and make a profit from, which was not the case of tourism or logging, both of which required capital resources and technical knowledge. The fact that agriculture is the most commonly accessible economic opportunity in frontier contexts is well established (Sunderlin et al. 2005). Thus, the surge of the agricultural trade once the road arrived would give previously off-road communities the cash they needed to send their children to school well equipped, properly uniformed and better fed.

Furthermore, the logging business, the main economic activity in the area, would also benefit from the reduction in costs the road would generate (INRENA 2008). Those involved in it, like the Andean council members of Diamante, tended to play down this possibility and its environmental costs, and some, like young Duncan, wanted illegal forestry in their territories to stop. The former insisted that this would happen automatically after the road arrived through the replacement of logging with agriculture. Other community leaders and authorities supported the same logic: timber was currently the only profitable business given the high cost of river transportation, but once the road was built agriculture would be made profitable, which would motivate people to leave timber behind and dedicate themselves to banana farming. Following this logic, the arrival of the road would reduce logging, an activity that was seen as the most significant threat that SERNANP and conservation organizations perceived (MINAM 2016).

As mentioned by Luis Manuel, the restaurant owner from Boca Manu, the reduction in transportation costs was also expected to increase access for tourists, something that would finally provide the income from tourism promised to locals since the foundation of the MNP (SERNANP 2014). Increased competition from more easily accessible protected areas, like the Tambopata National Park close to Puerto Maldonado, had forced the dwindling of the tourism industry in Manu in the last couple of decades (Shepard et al. 2010). Most of the tourist groups left were composed of high-end tourists who paid a high price for boat transportation directly into the Park and did not spend any time at the communities on their way there. As mentioned by Luis Manuel, this took a toll on local businesses, 'We don't enjoy one lonely cent from tourism. All tourists are brought in by agencies from Cusco and all the profit stays there. There is nothing left for Boca Manu!' According to others, the road would make it easier for low-budget tourists to arrive at previously off-road communities, like it did in Salvación and occasionally in Shintuya. The lodging infrastructure built in them would not sit idle as it did until then.

The two-way nature of the access provided by the road also promised lower costs for supplying communities with goods brought from outside. In Boca Manu Eduardo spoke to Adán, one of the merchants there. He had recently arrived at the port with a boat full of supplies for the three local shops, which everyone who was nearby dutifully helped to unload (see Fig. 6.2). People's excitement at his boat's arrival shows the significance of Adan's efforts to overcome the difficulties of river transport to supply his community. 'It usually takes us five days to bring the supplies for our shops. I suppose that the road would allow us to reduce that time to two days or so', he said. People in off-road communities had to pay more than double the usual retail price for things like rice, pasta, sugar and salt. These were a constant necessity in most households, which can be considered a result of acculturation in the case of the Indigenous peoples of Manu (Oliart and Biffi 2010; Ráez Luna 2017). The expectation was that the road would increase income and reduce the cost of these products, improving the ability of frontier dwellers to buy them.

Therefore, the promise of economic connectivity announced more profitable commerce of local products, the long-awaited arrival of tourists and cheaper goods that would improve the quality of life. It was also supposed to motivate people to leave logging behind for other businesses that would become profitable and were less environmentally harmful. After considering these aspirations, we now move on to our analysis of the realizations of on-road community members, as well as the perspectives of key actors regarding the possibilities for their fulfilment.

Fig. 6.2 Unloading Adán's cargo boat

6.2.2 *Timber, the Only Profitable Market—Realizations of Economic Connectivity*

As seen in the accounts Eduardo gathered, the fulfilment of the promise of economic connectivity seemed farfetched when visiting Shipetiari, which was ironically mentioned as an example of success by authorities and off-road community members. According to Chief Magno, banana sales had not been as good as proponents of the road had assured. His disillusionment was evident. The price they were being paid for the few bananas they sold was low given the risk wholesalers faced of having their trailers stuck on the road. They were not coming as often as the Matsiguenkas of this community had expected. Even when they would not buy all that they were offered, the few eateries in Puerto Shipetiari were their best clients. Still, Shipetiari's yuccas and bananas competed against the dried corn and dehydrated potatoes the Andean owners of these establishments would often prefer. The case of Shintuya was different but still not a great success of economic connectivity. Ricky and Nidia's trailer was the only one coming consistently and offering fair prices. They were the only ones who had made the effort to establish the necessary relationships to buy bananas from the Harakmbut people of this community, and this had happened several decades after the road had arrived there.

It seemed the line where the on-road banana market stopped being profitable would be drawn at Santa Cruz until the road surface was improved and bridges were built. Driving further down the road was impractical for wholesalers like Mrs. Gumercinda, as the risk on the bridgeless muddy road was too high for them. One obstruction would fatally ruin their cash flow. And even if the road surface was improved, banana producers from now off-road communities would be in a considerably weak position to negotiate with banana wholesalers. The reproduction of colonial ethnic hierarchies in this frontier context, in which Indigenous Amazonians remain 'inferior' to Andean migrants, is a key element of this disadvantaged position (Ames 2010; Morel 2014; Ráez Luna 2017). This had the potential to be worsened by what Tsing calls 'collectors' disadvantage' by which urban wholesalers stand in a favourable negotiating position regarding rural producers or collectors (Tsing 2005, p. 185). The combination of these elements significantly increased the possibility for instances in which trade is conducted in unfair conditions, something that is also frequent near protected areas (Adams and Hutton 2007; Murray-Li 2014). Moreover, the potential saturation of the Cusco market with bananas from places better connected to this city, like Puerto Maldonado and the rainforests of Cusco itself, even Santa Cruz, reduced the chances of commercial success that the bananas of now off-road producers had. This is often the case in frontier communities still distant from urban centres (Sunderlin et al. 2005). So, the prospects of the banana market further downriver did not look as good as proponents of the road were promising. It was clear that the main element of the promise of economic connectivity had little potential to be fulfilled.

Additionally, even if the banana market got to thrive because of the road extension, the argument that it was going to replace logging as the main source of income did not seem realistic. The Matsiguenkas of Shipetiari were disappointed by their lack of banana sales, but the reduction in transportation costs and the high demand of hardwoods across the river in Puerto Shipetiari had satisfactorily boosted their sales. So, the reduction in transportation cost was not enough to make banana farming profitable but it was enough to make timber trade even more profitable than it had been until then. This was especially so given the potential of frontier roads to foster illegal economic activity (Harvey and Knox 2015; MINAM 2016; Tsing 2009). This was also clear in the testimonies Eduardo gathered in Shintuya, where the banana market had only begun to emerge after the community's timber was almost completely depleted. Edmundo Kendero did not have the hardwoods he needed to build his boat with, while banana sales had only recently begun to pick up. The coexistence of both activities was evident at Shintuya's port, where stacks of timber and rotten bananas, a usual remnant of the banana trade, were both present (see Fig. 6.3). The situation was different in Puerto Shipetiari, where only stacks of timber and not a single banana skin could be seen.

Eduardo's interview with Pedro Igreda, a timber mill owner, in Salvación further informed the practical consequences of the road extension regarding the timber business. He said: 'When I started in this business we would only sell *cuartones.*[4] We would cut down hardwoods half an hour away from town. Nowadays we need to drive further, about two hours on the road and then up the riverbeds, to get to trees worth working with. If the road goes further we will definitely benefit because our costs would drop again: we wouldn't need to drive as far up the riverbeds, we would be able to cut down better quality trees near the main road. Right now, the

[4]Blocks measuring approximately 1 × 0.5 × 8ft.

Fig. 6.3 A stack of timber and litter from the banana trade at the port in Shintuya

best stuff is coming from Edén, from Shipetiari. They're going further and further to get the best wood, the stuff that pays better'. Evidently, the little speed provided by the Manu Road was only profitable for loggers, which is usual in first roads in the Peruvian Amazon (Larsen 2016). As will be discussed later, the argument that agriculture would replace logging accommodated well to the environmental interests that opposed the road.

It was clear, then, that illegal logging had a high chance of remaining as the main economic activity in Manu, regardless of road connectivity. As mentioned in Chap. 3, the dependence on this trade is unlikely to generate benefits locally given its boom and bust nature (Barbier 2012; Larsen 2016). Most probably, it will continue to generate the abuse, resource depletion and dispossession of Indigenous peoples that Eduardo's informants from Shintuya and authors like Oliart and Biffi (2010) describe as 'business as usual' for logging in the ACR. The highly illegal way in which the timber trade is conducted throughout Peru is also likely to keep reproducing the unfairness that characterizes logging deals and working conditions further along the Manu Road, which would constantly cause conflict between loggers and Indigenous people (Global Witness 2019; Leal et al. 2015; Urrunaga et al. 2012, 2018). As mentioned previously, Daniel from the MAA was already noticing the low prices that timber traders would offer for Shipetiari's hardwoods. This was not causing conflict yet but is an example of the potential for unfair trade relations fostered by ethnic hierarchies (Murray-Li 2014; Ráez Luna 2017; Tsing 2005).

On the other hand, tourism, frequently perceived as innocuous and 'sustainable' (Brondo 2013), had not generated much benefit to on-road communities either. As Eduardo's account shows next, the state of this business in Shintuya resembled what he saw in off-road communities:

During my stay in a house facing the port entrance in Shintuya, I saw very few of the tour boats that passed by actually stopping. Even less of them would spend money or more than fifteen minutes in the community. Ignacio Maqueri was with me when one of them made a brief bathroom stop. He asked, "How are we going to live from tourism if the tourists go straight into the park without visiting the communities? Most of them just stop and pee in our port!"

Despite the road and several tourism attractions within their territory, Shintuya was not getting a significant benefit from tourism, which is the same issue off-road communities suffered. The situation in Shipetiari was uncertain as their recent confrontations with isolated groups[5] had kept tourists away from there ever since, despite improved access. Thus, it seemed that the problem with tourists skipping the communities and going straight into the MNP did not have much to do with road accessibility but with several other issues.

Nonetheless, the reduction in transportation costs for goods brought from Cusco through the road did allow people from Shintuya to buy more processed foods from shops, which were slightly more numerous and better stocked than those in off-road communities. However, the causes and consequences of the demand for imported goods are interesting to consider, especially in light of the evidence about the ill-effects of market connectivity in the frontier (Murray-Li 2014; Sunderlin et al. 2005; Tsing 2009). The comments made by Soledad, a local Harakmbut activist, revealed how access to commercial goods was more a dependence than a choice, caused by the depletion of the forest resources her people used to live off of. Her comment reveals the way in which Indigenous people in frontier areas around the world are practically forced into a capitalist consumption system after significant changes in the land they live on (Murray-Li 2014). The dependence that Soledad mentioned, at the same time, increased the need for cash, which remained scarcely available to them. On the other hand, for José, in charge of the health centre in Boca Manu, the increase in access to processed foods was detrimental to the health of communities, 'People here would traditionally eat yuccas, which are a complex carbohydrate. Now they buy rice, pasta and wheat flour, which are simple carbohydrates that are not as nutritious. The same thing happens with proteins. Fish and bushmeat are replaced by canned tuna or frozen chicken, most of which is brought from Cusco in awful conditions'. Thus, the consumption of processed foods would reduce the quality of the locally produced diet. Additionally, he, Soledad and many others warned about the effects of an increased supply of processed alcohol, which seemed to increase consumption. Hence, the road did make imported goods cheaper but this had considerable implications on the health of community members, apart from cultural implications for Indigenous communities.

As seen here, the access to markets granted by the Manu Road was not sufficient to completely fulfil the promise of economic connectivity, and the improvements that did have the potential to be fulfilled could also have detrimental consequences. The aspiration of an increased income from the banana trade, which would benefit

[5]The Mashco-Piro that had killed a community member in 2015, mentioned in Chap. 5 (García Delgado 2018).

all off-road community members equally, had very slight possibilities of fulfilment. On the other hand, the road was most likely to boost the timber trade, which was only truly beneficial to small local elites and outsiders, and had the potential to keep expanding the abuse and depletion that it had already generated elsewhere in Manu (Oliart and Biffi 2010). Despite how evident this was in places like Puerto Shipetiari and Shintuya, off-road communities still placed their hopes and dreams of profitable trade on the road, revealing the 'enchantment' of economic connectivity.

This chapter about the 'enchantment' of economic connectivity, has explored the aspirations of off-road communities in Manu and suggested that the only economic activity that is most likely to benefit significantly from road connectivity is illegal timber extraction. As we will explain in the chapter that follows, the implications of this are significant, as the efforts of the local elites that are most likely to profit from this, and those of their political allies, further entrench the promises of the Manu road and reinforce its 'enchantments' (Harvey and Knox 2012, 2015). Therefore, the next chapter explores these and other types of political influence upon the 'enchantments' of the Manu Road.

References

Adams WM, Hutton J (2007) People, parks and poverty: political ecology and biodiversity conservation. Conserv Soc 5(2):147–183

Ames P (2010) Desigualdad y Territorio en el Perú: una geografía jerarquizada. Argumentos 1

Barbier EB (2012) Scarcity, frontiers and development. Geogr J 178(2):110–122

Barrantes R, Fiestas J, Hopkins Á (2014) Evolución de la infraestructura de Transporte y Energía en la Amazonía Peruana. In: Barrantes R, Glave M (eds) Amazonía Peruana y Desarrollo Económico. GRADE, IEP, Lima, pp 109–160

Brondo K (2013) Land grab: green neoliberalism, gender, and Garifuna resistance in Honduras. University of Arizona Press, Tucson

Campbell JM (2012) Between the material and the figural road: the incompleteness of colonial geographies in Amazonia. Mobilities 7(4):481–500

Carvalho GO, Nepstad D, McGrath D, Diaz MDV, Santilli M, Barros AC (2002) Frontier expansion in the Amazon: balancing development and sustainability. Environ: Sci Policy Sustain Dev 44(3):34–44

Congreso de la República (2013) Proyecto de ley que Declara de Necesidad Pública e Interés Nacional la Construcción de la Carretera de Integración Regional de Madre de Dios. Comisión de Transportes y Comunicaciones, Lima

García Delgado F (2018) Comunero muere por impacto de flechas de indígenas mashco piro. El Comercio Website. https://elcomercio.pe/peru/madre-de-dios/contactados-matan-flechazos-comunero-peru-noticia-501491

Global Witness (2019) El justiciero forestal: Por qué se debe devolver la independencia a OSINFOR y expandir sus poderes. Global Witness, London

Harvey P, Knox H (2012) The enchantments of infrastructure. Mobilities 7(4):521–536

Harvey P, Knox H (2015) Roads: an anthropology of infrastructure and expertise. Expertise, Ithaca, NY

INRENA (2008) Plan Maestro de la Reserva Comunal Amarakaeri 2008–2012. Ministerio de Agricultura, Lima

Jarrier A (2012) Seminario Internacional: Todos los caminos conducen al Brasil. La geopolítica del plan de Integración de Infraestructura Regional Suramericana—IIRSA—en la región andina. In: Bulletin de l'Institut français d'études andines, vol 41 pp 303–307
Larsen PB (2016) Derechos indígenas, gobernanza ambiental y recursos en la Amazonía peruana, Hacia una antropología de la posfrontera. Instituto de Estudios Peruanos, Lima
Leal D, Salisbury D, Faquín Fernandez J, Cauper Pezo L, Silva J (2015) Ideas cambiantes sobre territorio, recursos y redes políticas en la Amazonía indígena: un estudio de caso sobre Perú. J Latin Am Geogr 14(2):181–204
MINAM (2016) Dossier: ¿Por qué no debe aprobarse el predictamen de insistencia de la ley sobre la carretera Madre de Dios, tramo Nuevo Edén—Boca Manu—Boca Colorado? Ministerio del Ambiente, Lima
Morel J (2014) De una a muchas Amazonías: Los Discursos sobre 'La Selva' (1963–2012). In: Barrantes R, Glave M (eds) Amazonía Peruana y Desarrollo Económico. GRADE, IEP, Lima, pp 21–46
Murray-Li T (2014) Land's end: capitalist relations on an indigenous frontier. Duke University Press, Durham and London
Oliart P, Biffi V (2010) Territorialidad indígena, conservación y desarrollo: Discursos sobre la biodiversidad en la Amazonía peruana. Instituto del Bien Común, Lima
Pieck SK (2011) Beyond postdevelopment: civic responses to regional integration in the Amazon. J Cult Geogr 28(1):179–202
Quinhoes RC (2011) Redes y territorio: la iniciativa IIRSA en foco. Espacio y Desarrollo 23:5–29
Ráez Luna E (2017) Derechos Ancestrales y Conservación de la Naturaleza en Debate: El caso de los Pueblos Indígenas del Parque Nacional del Manu In: Ráez E (ed). Seminario Permanente de Investigación Agraria (SEPIA), Lima
SERNANP (2014) Plan Maestro del Parque Nacional del Manu 2013–2018. Ministerio del Ambiente, Peru
Shepard GH, Rummenhoeller K, Ohl-Schacherer J, Yu DW (2010) Trouble in paradise: indigenous populations, anthropological policies, and biodiversity conservation in Manu National Park, Peru. J Sustain Forest 29(2–4):252–301
Sunderlin WD, Belcher B, Santoso L, Angelsen A, Burgers P, Nasi R, Wunder S (2005) Livelihoods, forests, and conservation in developing countries: an overview. World Dev 33(9):1383–1402
Tsing AL (2005) Friction: an ethnography of global connection. Princeton University Press, New Jersey
Tsing AL (2009) How to make resources in order to destroy them (and Then save them?) on the Salvage frontier. In: Histories of the future, pp 51–73
Urrunaga J, Johnson A, Orbegozo D (2018) El Momento de la Verdad: Oportunidad o Amenaza para la Amazonía Peruana en la lucha contra el Comercio de la Madera Ilegal. Environmental Investigation Agency, London and Washington, DC
Urrunaga J, Johnson A, Orbegoso I, Mulligan F (2012) La máquina lavadora: Cómo el Fraude y la Corrupción en el sistema de concesiones están Destruyendo el Futuro de los bosques del Perú. Environmental Investigation Agency, London and Washington, DC

Chapter 7
Territoriality and Power in Manu

Abstract In this chapter we analyse the political forces that underlie and reinforce the 'enchantments of infrastructure' in Manu. First we present Eduardo's ethnographic accounts, which provide insights into how logging elites and their political allies use their power to influence the local narrative surrounding the road and the expectations of its environmental consequences. This is followed by a section in which we analyse the influence of national conservation strategies, policies affecting Indigenous territories and neoliberal infrastructural policies on the weakening of Indigenous territoriality. We argue that the combined influence of these elements ultimately reinforces the power of the 'enchantments' of the Manu Road. We conclude this chapter by presenting a section that exemplifies the way in which this blend of elements and its effects are evident in local discourses about the issues surrounding road construction. Our previous evaluation of the 'enchantments of infrastructure' is complemented by the analysis of these underlying aspects to lead to our concluding remarks in the following and final chapter.

Keywords Manu road · First roads · Frontier infrastructure · Enchantments · Indigenous · Amazonia · Political ecology · Conservation · Territoriality · Power · Conflict

7.1 Stories of Power

The following accounts explore issues around power relations and political manoeuvring that became evident during Eduardo's fieldwork in Manu. As our analysis shows, these elements have an important effect on the capacity of the Manu Road to 'enchant' off-road communities.

E. Salazar Moreira and M. Palomino-Schalscha, *Road Expansion in the Peruvian Amazon*, SpringerBriefs in Latin American Studies, https://doi.org/10.1007/978-3-030-47182-8_7

7.1.1 *"I Am the Chief Now!"—Power Relations in Diamante*

After being in Diamante for a few days, I was invited to participate in the community's Mother's day celebrations at the community centre. There I was able to see all community members gathered in one place. There were a couple of hundred attendants, the great majority of them Yine, and only a few Andean people; the schoolteachers and those who had married into the community. Duncan, as the council's secretary, was the only Yine person co-leading the celebration and wishing a happy Mother's Day to those women who didn't speak much Spanish. My baskets were given out to two Yine mothers who seemed grateful to "Mr. Endimburo, from the University", as the signs on them said. Carmen constantly repeated that the baskets had food and materials worth 50 soles,[1] and that she had managed to get all the donations needed for them.

The party was a great chance to meet the local families and introduce myself informally. What surprised me was that the people that seemed to be honoured, by the community's mothers instead of the other way around, were the Andean community members, the teachers and myself. We were taken around the community centre where mothers were sitting and giving us cups of their *masato*, a home-made yucca beer famous throughout the Peruvian Amazon.

I hid my reluctance to participate in the dance contest. Robert's wife was nice enough to choose me as a partner, most of the Yine men had already left. Only Andean men remained, drinking more expensive bottled beers among themselves. The mothers who won were given *masato* in 2.5 or 3 litre plastic bottles, which they were supposed to drink in one go as they danced to another tune. After helping Robert's wife win one of the dancing rounds, I excused myself with her to avoid the several more offerings of *masato* I was sure to get. If there was one thing that was clear to me after leaving the party was that, despite the low ratio of Andean migrants, they seemed to be receiving all the honours and making most of the decisions for the festivity.

The dominance of Andean migrants was also notorious in other occasions. After two days of celebrations, I was happy to return to what seemed to be normalcy in Diamante. By that time, I had had informal conversations with several people from the community and was eager to formally interview someone, recording and making notes. Carmen organized a group interview with the community council, I was not allowed to interview them individually.

Moments before the group interview, Mr. Axel, an Andean man who was the community's treasurer, took the liberty of talking to me while Carmen was still absent. He spoke about the exclusive nature of logging in the community. 'Only those of us who have the means, the skills and the money to do it work with timber. If we had a road, then everyone could benefit because those unable to work with timber would be able to sell the products they harvest. Everyone knows how to farm here'.

---

[1] About USD$ 15.00 at the time.

Once the group interview started, he, Celia and Carmen, who clearly monopolized the conversation over the other two council members, who were Yine, kept explaining this argument. 'Timber is the only thing that allows people to make a profit here. Those who can't do it live without anything. If the road comes they would be able to work with bananas, with yucca. Not only that, everything grows around here! But without a road, how are they going to sell it? With the road we might even forget what timber is!', is what Axel said, followed by a loud 'Of course!' from Celia. Carmen continued, 'As he says, if the road passes by my community we would leave timber behind a bit because we would work with agriculture. Not all community members have the tools required for logging but everyone has farming tools. You've got a machete, you've got an axe, so you can work the land. A hectare or two, you've got what you need to produce. For logging you need chainsaws and all-terrain-vehicles, you need lots of things!' Celia followed by reinforcing their point by putting forth the notion that this would nullify the potential environmental impact of the road. My conclusion, however, regarded the return to the investment loggers had made to start their business before the road arrived. Why would they stop making a profit from logging, especially after investing in all this expensive equipment? Several government authorities and prominent proponents of the road would also mention this argument, which took a life of its own.

After the interview, I hung around Celia's shop and the council started their own meeting while I finished my notes. It seemed the chiefdom proceedings were ongoing despite Carmen's assurance to me that she was the Chief. She began by telling them, 'We need to replace the Chief quickly to keep moving forward. We cannot elect a new directive council because we will waste time. If we do we won't be able to do anything, if we wait for Jorge. If I assume chiefdom, we will get things done quicker'.

Their meeting went on and I finished my notes. These featured frequent jottings about how Carmen, Celia and Axel constantly interrupted the only Yine woman in the group, Mina, during the whole interview and the meeting. Despite being the appointed spokeswoman, she only spoke in a weak voice and was hardly allowed to finish her sentences before others spoke. Duncan only agreed to or paraphrased what the Andean council members said.

Disregard towards Yines was also noteworthy in more private occasions. Early one morning I was interviewing Nehemías, a local fisherman, at Celia's shop. He was explaining his views on the Yine-speaking isolated groups that had recently been involved in violent encounters in Shipetiari, a neighbouring Matsiguenka community, and drinking *masato*. 'I have *masato* for breakfast usually. Sometimes we can't afford food so we fill our stomachs with *masato*, which keeps better than raw yuccas. That's one of our few options to keep our heads up for the day's work'.

As I spoke to him, Carmen and her good friend and neighbour Neptalí, the former community Chief, who was also of Andean origin, were eating the fish Nehemías had sold them. Apparently, there was something wrong with it.

'Nehemías, you motherfucker! You sold us rotting fish!', said Neptalí.

'Hey, Nehemías, I swear I'm going to kick the shit out of you!', complemented Carmen.

As Nehemías shrugged it off and laughed, I assumed this type of talk to be a usual joke between them. What surprised me was that they both made it while I was talking to Nehemías. By then it was clear that the Andean residents of Diamante wanted to hide certain things from me but their contempt towards their Yine neighbours was not one of them.

Something that added to this suspicion was how Carmen and Neptalí dealt with the meetings the SRG was holding with the leadership councils of the various communities to discuss the road project. There was one in Boca Manu while I was still in Diamante. Neptalí brought the news to Carmen and they went together. Duncan did not know that they had gone, despite being part of the council as well, and didn't act surprised when he knew they had gone without notice. The Vice-chief and the ex-Chief, both Andean migrants, had gone by themselves to represent the Yine community at a meeting with the regional authorities. No Yine people were representing the community.

Weeks before my visit to Diamante, Edmundo, from Shintuya, had told me that it was usual for Neptalí to do this. I was sceptical then. Sometime later, in Reinalandia, I interviewed Miki, a Matsiguenka boat driver who had been living in Diamante for several years after marrying a local woman. His comments about Carmen and Neptalí made things clearer. 'The way I see it, they manage things to their own benefit, they don't share the management with the community. That's just for them. They decide things themselves. They call the community for a meeting and say, "This is what we have decided, now you sign the papers." Nothing more, they don't share the management. Now they're still the chiefs of the native community. Carmen is Neptalí's best friend. If someone from the community is elected for the council they only say they're incompetent, that they're no good'. It seemed Miki had a more critical view of things, maybe because of his extensive travels through the southern Peruvian Amazon as a tour boat driver. Furthermore, considering that Carmen and Neptalí were married to Yine siblings, and that they interacted constantly at Celia's shop during my stay, I began suspecting that Carmen was just a stand-in for the former Chief.

Definitive proof of this came on the day I left Diamante. Early in the morning, a group of representatives from the SRG arrived and gathered at Celia's shop. They were there to agree on the terms under which they were going to conduct the studies required by the road project within the community. Neptalí came out to greet them wearing an ACR t-shirt and vest, very official attire. Carmen stepped back and let him speak to them.

Neptalí was the de facto chief, at least for dealings regarding the road. It appeared that they had strategically used a seemingly administrative argument to get rid of Jorge, the community's first Yine Chief in a long time.

7.1.2 "We Will All Become Farmers"—A Puzzling Argument Emerges

The apparent panacea that the Manu Road appeared to be met a wall of opposition from SERNANP, and its parent organization, the MINAM, as well as the many conservation NGOs that worked in the area. Luckily for me, the organization I had worked for before had not signed a controversial letter of opposition to the road back in 2015. Apart from working with the National Park, we worked with local farmers who supported the project, so signing it would have been hypocritical from our part and risky for our relations. Given the very likely environmental consequences, the letter was categorical in its rejection of the road project and GOREMAD's intent to go forward with it. This without much recognition of local needs. The text was followed by a long list of adhering organizations. Some of them most locals had not even heard of, and a few others had earned a reputation as foreign usurpers after being given large private conservation concessions adjoining the MNP and the ACR (see Map 2.1).

'You're still our friend. Of course you're welcome in my office! Your guys were one of the few NGOs that supported our efforts for the road', was how Ricardo Gómez, manager of the Manu SRG, greeted me at his office in Salvación. He was alluding to the infamous letter, which had clearly marked who was on either side of the argument. I was nowhere as sure of my own position as he was. His contempt for his antagonists was even stronger for what happened after the letter. He, his boss, Governor Luis Otsuka,[2] and the SGR team of engineers had been sued by MINAM in late 2015 for building the road from Nuevo Edén without the approval they needed, which required an environmental assessment process they allegedly could not afford.

He clearly took this as heroic martyrdom, as he started explaining the benefits this had already generated for locals as soon as he finished asserting his innocence. 'Building that local track from Nuevo Edén to Puerto Shipetiari and beyond cost us much. But, look, just by building that 10 km stretch now we have an agricultural fair in Itahuanía. No one would have thought that the area beyond Shintuya would be having such an event! Most people there worked with timber exclusively but now that they have road access, they have started dedicating their time to growing more bananas. That is why we are running the agricultural fair there for the second year now!' It certainly surprised me to see banners for the Itahuanía agricultural fair across the main road in Salvación. A *cumbia* band would be playing, which made the celebration official.

The process that Gómez suggested as having already happened in Itahuanía and communities further down the road had also been mentioned as a given by the community council in Diamante. I would hear it repeatedly from local leaders and authorities throughout my fieldwork: Logging was the only viable business for off-road communities given the high cost of river transportation, which was not affordable through banana sales. Proponents of the road assured that when off-road communities became

---

[2]This name has not been changed for a pseudonym as former Governor Luis Otsuka is a public figure and was not an informant for this research.

connected to the road they would be able to sell the bananas they produced to wholesalers that supplied the market in the city of Cusco at a cheaper cost. This would give these communities an alternative to timber extraction, and they would all switch to farming. This also seemed to be the hope of young Duncan, from Diamante, 'Little by little we're running out of trees. We want to be able to protect our forest. We don't want people to keep logging, we don't want them to cut timber anymore. The community will sell more bananas and work the land to sell their produce once the road comes'. This logic seemed to fit well with the environmental concerns raised by the conservation organizations that opposed the road. I opened myself up to the possibility of this seemingly promising change.

However, every time the argument was mentioned, I couldn't avoid finding contradictions. Gómez's example regarding the Itahuanía agricultural fair was the first to seem flawed. I had never heard contrasting points of view regarding Nuevo Edén and Itahuanía during my two years of work in Manu. As he had said, these communities were regarded by most as logging settlements. Their representatives at meetings with the SRG or the Municipal Government would always support initiatives that benefitted logging. These included the reduction of the ACR's buffer zone and other measures that were severely controversial with the few more vocal Indigenous representatives, like my friend Soledad from Shintuya.

More proof would come later. In one of my outings along the existing road, I rode on a pickup truck with some other travellers on my way to Puerto Shipetiari. Mr. Castro, community President of Nuevo Edén, who was on the front passenger's seat, constantly referred to himself and his community as 'us farmers' while speaking of this supposedly immediate change towards agriculture. Castro supported the road quite strongly. He assured that 'there would be trouble' if it was not built quickly, 'Just like there was in my homeland San Martín,[3] where they were trying to stop the construction of the road that goes through the Alto Mayo.[4] They didn't want to allow the road until the communities burned a police car'. He sounded ready to pull out a matchbox.

It was raining heavily and when the driver decided to pick up an elderly man who was walking by the side of the road, I decided to give him my seat inside the car. I sat on the car's cargo bed next to Ismael Kamene, a Harakmbut man who had hopped on the car in Shintuya. I knew him from before. Looking out onto the road from the back of the car I noticed that the forest on the sides of the road seemed much taller when we came closer to Itahuanía. Ismael explained, 'That's because no one in these towns clears the forest for farming, they just work with timber here. They cut down the hardwoods and leave the lower quality trees standing. Once they run out of hardwoods I'm sure they'll start cutting the cheaper trees'. There were no signs of farming plots beyond Shintuya and logging tracks and stacks of timber on the sides of the road began to abound.

When we passed by Itahuanía, I saw the only public display in favour of the road I would see during my fieldwork. A sign that said, 'Mr. President Ollanta Humala

[3]The San Martín region of the central Peruvian Amazon.

[4]The Alto Mayo Protected Forest, *Bosque de Protección Alto Mayo*.

Picture 7.1 The sign in support of the road in Itahuanía

Tasso, the people of Manu greet you. We want road integration between Manu and Puerto Maldonado! Nuevo Edén–Boca Manu–Boca Colorado' (see Picture 7.1). The sign had been made especially for a highly expected visit from the Peruvian President back in 2015. I remembered the excitement that everyone felt in that occasion. People flocked to the bridge over the Carbón river,[5] the last one on the road before it entered the Madre de Dios region. Despite his apparent military toughness, Humala never arrived. The road from Cusco had proven too risky for the mandatary.

I kept wondering about the sign though. They had kept it up for so long for everyone to see. Why were on-road communities supposedly dedicated to banana farming still demanding the road extension so actively? If the target market for their bananas was the city of Cusco, to where they already had road access, then further extension of the road should not have had much value for them. Unless they were unusually supportive of off-road communities or were interested in increasing their own access to timber stocks further down the AMD river. This sign was a testimony of the interest of the illegal timber trade in the expansion of the Manu road.

When talking to Ricky and Nidia, the banana wholesalers I knew from Salvación, I had proof of the latter. 'We have gone there recently to see how things went with them. We thought we could try to start trading with them like we do in Shintuya. They had planted the banana suckers that the [Agricultural] Agency had given them. Their fields were near their settlement and racemes hung from their plants. But they wouldn't harvest them for us, we had to go get them ourselves. Of course, we haven't gone back. They only think of timber because that gives them a lot of money

[5]This was the third rendering of the bridge over the Carbón. The first two had fallen into the river as they had been built precariously. The current version stood strong but there was a corruption scandal surrounding its construction.

very quickly', said Nidia. Several things seemed to indicate that the transition to agriculture had not taken place in those communities at all, despite the road.

More examples to support this argument were given by other advocates of the road project. *Profesor* Fernando Paniagua provided an interesting comparison. 'I have seen what happened in Palotoa. Palotoa doesn't work with timber any more, just bananas, and that is because of the road. Right now we don't see farms here but I am sure that when the road comes, they will start farming within a year. I don't think they will keep logging. They are logging now because there is no other job available. So, they at least do it secretly, as contraband. They are forced to do that to survive, to make a little profit, at least to have a drink to forget about this situation. With the road people will start producing, they will have better lives. Young people who study at the school here will be parents by then and they'll be able to educate their children in the cities'. Paniagua had been a teacher in Palotoa Llactapampa in the 1980s, so his comparison of the past situation there and what he thought would happen once the road connected off-road communities seemed believable.

However, when he explained the process that Palotoa had gone through, this argument seemed flawed again. He wasn't portraying Palotoa as a farming community when it was not, as Gómez had done with Itahuanía. Palotoa Llactapampa was regarded by everyone in Manu as the example when it comes to banana commerce. What didn't appear to fit was the very motivation for the switch to agriculture, 'They didn't do anything more than cutting timber before the road to Cusco was improved in the 90s. When the road was bad they busted all their timber to survive. When they ran out of it they starved for some time until the truck that came to Santa Cruz to sell them grain and vegetables from Cusco started buying loose banana racemes from them. Once that started happening, people were motivated to produce more bananas and started a constant supply there'. So, the banana trade in Palotoa hadn't started until well after the road was improved and they had run out of timber already. The change to farming was not an immediate consequence of road connectivity but a long-term aftereffect of the depletion of timber stocks.

Profesor Paniagua had explained the dire consequences he and his staff had to go through to collect and cash their cheques in Salvación and visit their families in the cities. It seemed to me that these were Paniagua's real motivations to want the road, and understandably so. Why had the headmaster adopted this flawed argument?

People who were permanent residents of Boca Manu provided a different description of what they thought would happen when the road reached them. Silvia, the shop owner, made a quick comment that showed how flawed the logic proposed by Gómez, Paniagua and several others was, 'I haven't cut any timber from my land. Why would you cut it if your profit was still going to be low? I'll do it when the road comes. That's going to be better'. Adán, another one of the merchants from Boca Manu, was unusually open to me about his suspicions regarding the road and the SRG's involvement. 'I don't think much will change. When the road comes the government will still support those that work with timber'. He had been a boat driver for tourism agencies in Manu for a long time and had personally witnessed the 'boom and bust' nature of logging as it advanced with each extension of the road and how only the 'bust' would give way to agriculture. In his eyes, agriculture was not promising either. 'I wish everything was focused on tourism but it's all about timber here. The road comes and the timber goes, that's what happens everywhere. In the first few years there will be trucks moving timber but they will finish off the little that's left fairly quickly. They'll keep logging anyway but not as much. Some will gravitate towards agriculture, others will see how that won't help them and move somewhere else. If people move on to agriculture, the transportation cost through the road will still be high. Probably the farthest from Cusco that agriculture might be worthwhile is Shintuya, maybe Itahuanía, because they are closer. If the road from Paucartambo[6] was in better condition, transport costs would be cheaper. Fruit also rots too quickly to make the trip'.

Ryan from Shipetiari had similar expectations of what would happen, 'If the road continues they, the migrant settlers, the loggers, are going to keep going further in. Moving forward after the timber'. On my way back from Shipetiari, I took some more time to explore the port across the river and saw what Ryan meant. While I waited for my transport, I walked up the side road that led to the main road from the port. Along it, I found an unusual boatyard. Blue sparks flew off a blowtorch. Its operator explained, 'We've been making steel boats for a while here. It's much easier to get steel plates from Cusco now than it is to get the right kind of wood around here. They're running out of good timber'. This surprising fact made me think of Edmundo's lack of a boat; would he ever make enough money to buy a steel boat?

I kept walking up the side road, having a curious look around. When I arrived at the fork on the road, I saw three *trocheros* parked, seemingly awaiting orders. The narrow section of the road that followed on but didn't get to Diamante was covered in their tracks. These continued from the fork further into the forest, just like the logging tracks that would come out perpendicularly from the road near Itahuanía and Nuevo Edén. If there were no towns connected by the road after Puerto Shipetiari, the only reason for these vehicles driving into it was penetrating the forest. Later, in Salvación, Roger Garibaldi, who operated an adventure tourism business, would tell me what he had seen when he drove into that stretch of the road: 'That's primary forest there. There are big trees right at the edge of the road and you can see spider monkeys on them. But it's full of *trochero* tracks and litter from the loggers. They're cutting down trees at the edge of the road'.

After having looked around Puerto Shipetiari for hours while I was stuck there, I gave up. There was nothing, not a single raceme stem, not even a banana skin, was to be seen in the whole port. These are the tell-tale signs of a banana trading spot: raceme stems, abandoned racemes that were too ripe to make the trip to Cusco, and the sweet smell of rotting bananas. I didn't see any of that among the stacks of timber in Puerto Shipetiari (see Picture 7.2). In contrast with Shintuya (see Chap. 6), the banana trade was nonexistent in this port, established immediately after the last road expansion. So, the current use of that new stretch of road was clear. The nonexistence of the banana market and the continuity of the illegal timber trade in Shipetiari were

[6]Paucartambo is the last large town the road from Cusco passes through before descending from the Andean highlands into the cloud forest and the Manu rainforest below. The tarmacked section of the road ends there, the remainder of the way to Manu is a dirt track.

Picture 7.2 A stack of timber in Puerto Shipetiari

apparent. However, it didn't seem that the Matsiguenkas were benefitting much from the port itself. Ryan had told me that the plot they had been given by the Nuevo Edén leadership to build a communal storage house on had been 'swallowed' by the river. They had been given a piece of land vulnerable to that, a common effect of the river's meandering that had also done away with the police station in Boca Manu. They didn't have anywhere in the port to store their products for sale or their gear. The timber trade continued and the Indigenous communities continued to benefit little from it.

7.2 Unpacking Territoriality and Power on the Manu Road

Having uncovered the way in which the 'enchantments of infrastructure' proposed by Harvey and Knox (2012) manifest themselves in the case of the extension of the first road in Manu, we now turn to exploring the political forces that affect them. In the next section, we analyse the way in which the establishment of the MNP and the ACR have constituted acts of territorialization and generated a sentiment of rejection in the great majority of the people of Manu. We continue with a section that analyses how the form of territorialization reproduced by the Manu Road fits well with local aspirations for modernization, to the point of facilitating the silencing of contesting voices. Our analysis of territorial issues concludes with a brief section explaining how the combined effects of the consequences of these issues reinforce the 'enchantments of infrastructure' in Manu.

The powerful actors that promote modernizing development through the construction of roads that serve extraction and those that promote environmental conservation through the establishment of protected areas enter in conflict with each other over land control in Manu. Despite their apparent opposition, they do so through similar means, as both political forces have established structures that have weakened the territoriality of the Indigenous peoples of Manu to achieve their goals (Baletti 2012; Oliart and Biffi 2010; Ráez Luna 2017). The creation of the Native Community framework confined Indigenous territory by pressuring for sedentarization, which left the land that Indigenous peoples were not able to claim or effectively protect available for extraction (Larsen 2016; Leal et al. 2015; Peluso and Alexiades 2005). The establishment of protected areas, especially the MNP, ignored, disenfranchised and dispossessed Indigenous peoples settled previously in the area through the top-down administration of the land by the State's conservation agencies (Adams and Hutton 2007; Larsen 2016; Oliart and Biffi 2010; Ráez Luna 2017; Raycraft 2018; Shepard et al. 2010). Thus, these forces have turned Manu and other places in the Peruvian Amazon into territorialized spaces that are controlled by and generate the most value for outside interests (Peluso and Lund 2011). We propose that the extension of the Manu Road is, in several ways, the resolution to this multi-lateral territorial confinement that local communities see as most favourable, which reinforces the power that the road's 'enchantments' would usually have over frontier areas.

7.2.1 "Like Monkeys in a Cage"—Territorialization for Conservation in Manu

For anyone that has spent time in Manu away from the friendly smiles of their tour-guides and SERNANP's park rangers, the animosity most local people feel towards 'the Park'[7] will have become clear. Most local farmers, Indigenous people not affiliated to extra-communal organizations, government officials and especially elected authorities and loggers will harshly complain against the MNP and the ACR. The opposition of MINAM, SERNANP and conservation NGOs towards the Manu Road project, given the threat it poses to the preservation of the MNP's ecosystems, has focalized the popular estrangement of the communities in Manu towards conservation (Gallice et al. 2017; MINAM 2016). The most regular judgement towards SERNANP and conservation interests that was manifested by pro-road informants in Manu was quite negative. Luis Manuel, the restaurant owner from Boca Manu, said, 'They don't care about the needs or the hunger of Native Communities or Boca Manu. They only care about filling their pockets, becoming rich themselves and keeping us like monkeys in a cage'.

The source of this contempt can be traced back to the restrictions that protected areas in Manu represent for of the province's residents (Oliart and Biffi 2010; Ráez Luna 2017; Shepard et al. 2010). This is not limited to the lack of road connectivity

[7]'*El Parque*' is the way most locals refer to the MNP's management.

that the organizations that manage these protected areas have sought to maintain but also to the limitations to resource use that these have historically attempted to enforce without providing any viable alternative (Gallice et al. 2017; Oliart and Biffi 2010; Ráez Luna 2017; Shepard et al. 2010). The economic cost that Peruvian protected areas represent for the communities under their influence is evident along the Manu Road and the AMD river, especially in off-road communities (Barrantes and Glave 2014; Díaz and Miranda 2014; Oliart and Biffi 2010; Ráez Luna 2017). The accounts that we have presented along with this work, especially the testimonies from Isla de los Valles and Diamante, are examples of this situation and of the perception that the Manu Road would finally end this condition. Comments made by Celia, the council member from Diamante, also reflect this, 'We suspect of the Park because, why would they oppose the construction of the road if it goes through our land and not theirs? What do they give us in exchange for our contribution to that area? It would be good if they would tell us they were going to improve our water or give us communication, 'I'll help you, I'll build something for you.' But they don't. Our school is pitiful. Our senior high school students don't know how to work a computer. They go to the city and they are null. If the Park would solve that situation then they'd have a reason to say "no" to the road'. *Profesor* Paniagua's views went along the same lines, 'The State prosecutes people instead of guiding them in their use of resources. Environmental regulations have to be enforced to protect nature but that doesn't mean they have to abandon the people that live here. They're really good with restrictions but when it comes to helping communities change their livelihoods for something else they don't help at all! That's why the road is an absolute necessity for those who live here. That is why the road always signifies development'. Perceptions like these are also an example of the expectation that communities influenced by protected areas have for compensation to the costs they incur in for the sake of conservation, as well as of how infrequently this is provided (Kohler and Brondizio 2016).

Despite any claims and expectations from affected communities, the perception of conservation towards local resource users as despoilers of valuable ecosystems justifies their exclusion from the management of protected areas, and from the areas as physical spaces as well (Adams and Hutton 2007; Bluwstein and Lund 2018; Peluso and Lund 2011; Ráez Luna 2017; Raycraft 2018; Tsing 2005). This exclusion is certainly perceived by communities in Manu and their leaders, who identify it as a result of the imposition of the will of foreign actors (Oliart and Biffi 2010). This generates a high degree of animosity, which can be seen in comments by Mr. Castro, from Nuevo Edén. He told me, 'As citizens of Manu province we can't go inside the [Manu] National Park. SERNANP always lies about the situation here. They never say what the population says about the issue, they change it. When we brought people from Tayakome and Yomibato[8] to the strike for the road in Boca Manu they told them what they thought. Those people also want the road but the Park didn't tell that to anyone'. On the ground this sentiment can also be easily noticed at the 'participatory' meetings organized by SERNANP and MNP or ACR authorities, which often have a tenor of confrontation instead of participation. Eduardo has personally witnessed

[8]The two Native Communities inside the MNP (Ráez Luna 2017).

several occasions in which community leaders and local authorities have verbally bashed SERNANP officials, who often take this type of abuse heroically and without any reasonable retort. Thus, SERNANP and conservation organizations have little leverage with local communities regarding the road debate, a controversy which has rekindled dilemmas left unsolved in the past (Oliart and Biffi 2010; Ráez Luna 2017). This often makes further exclusion and regulation the only means that this national government agency has to uphold the protection of the MNP and the ACR, which is a weak resource when confronted with popular opposition (Adams and Hutton 2007; Bluwstein and Lund 2018; Díaz and Miranda 2014; Larsen 2016; Oliart and Biffi 2010; Ráez Luna 2017; Robbins 2004; Tsing 2005). The minutes of meetings focused on the road project provide examples of this, as the answers of SERNANP's representatives mostly reminded local participants of the power this agency had to veto the road project by not providing their required approval (MINCUL 2016a, b, c). The weakness of these measures is clearly shown by SERNANP's later approval of the road project and by testimonies Eduardo gathered from park guards, manifesting that popular will in Manu had left little choice but for the MNP to adapt its protective measures to the existence of the road (SERNANP 2017).

Thus, this process reveals the long-term implausibility of sustaining restrictive conservation measures based on the exclusion of Indigenous peoples, other local communities and their will (Gallice et al. 2017; Oliart and Biffi 2010; Ráez Luna 2017; Shepard et al. 2010). The lack of participation of local communities translates into a lack of 'ownership' of the protected areas that surround them, which turns the institutions that manage these spaces into their antagonists (Oliart and Biffi 2010). Furthermore, the negative economic effects that Peruvian protected areas have on households under their influence, and the fact that these are largely ignored, eventually becomes a powerful source of pressure upon conservation (Díaz and Miranda 2014; Kohler and Brondizio 2016; Larsen 2016). This ultimately shows how attempts to isolate valuable ecosystems from the human activities that take place in them through restrictive measures often result in ongoing conflict (Larsen 2016; Robbins 2004). Said conflict, which in Manu manifests itself through the road controversy, leaves a little option to conservation institutions in the long term, which compels these to concede to pressures that have intensified with time (Kohler and Brondizio 2016; Larsen 2016; Ráez Luna 2017; Shepard et al. 2010).

In this section, we have established the unsustainability of exclusionary conservation in Manu by providing evidence of its effects and examples of local sentiment towards it (Gallice et al. 2017; Oliart and Biffi 2010; Ráez Luna 2017; Shepard et al. 2010). We now turn to our analysis of the local reaction towards the type of territorialization reproduced by the Manu Road.

7.2.2 The Magic of 'Neoliberal Territorial Design'—Conjuring and 'Enchantments' in Manu

The type of territorialization generated by road building in Manu, on the other hand, plays out more favourably for the forces behind it. The neoliberal territorial design promoted by IIRSA, and furthered by the smaller road projects that stem 'silently' from it, promises economic growth through market connectivity (González 2012; Jarrier 2012; Kanai 2016; Zibechi 2006). This makes neoliberal territorial design compatible with the aspirations of modernization held by frontier communities, reinforcing the 'enchantments of infrastructure' (Harvey and Knox 2012; Kanai 2016). Thus, despite the potential environmental and social risks that the Manu Road forebodes, its appearance as a modernizing infrastructure that will help communities generate the economic growth they need to survive in the global economy turns it into a welcomed territorial intervention (Gallice et al. 2017; Harvey and Knox 2012, 2015; Larrea-Gallegos et al. 2016). This discourse of 'the road as development' is noticeable in several of the ethnographic accounts presented so far, where local authorities and off-road community members speak of this infrastructure as the solution to most of their problems. However, the illusory nature of this is also evident in the account described in Sect. 7.1.2, as the effects presented as the proof points of the Manu Road, that everyone would leave logging behind to become farmers, seem not to have taken place as represented by its proponents (Harvey et al. 2017; Harvey and Knox 2012, 2015).

In several ways, the stage is set for the type of territorialization that road networks promote in the Peruvian Amazon. In Peru, the neoliberal territorial design perpetuated by the Interoceanic Highway, as a component of IIRSA, is preceded by the creation of the Native Community framework (Kanai 2016; Leal et al. 2015; Peluso and Alexiades 2005). Through the imposition of this framework, Indigenous communities become sedentarized and their territories confined into smaller units than what their traditional livelihoods would have required (Larsen 2016; Salisbury et al. 2011). Their displacement from larger territories leaves land free for the type of resource extraction that IIRSA seeks to facilitate (Kanai 2016; Leal et al. 2015; Peluso and Alexiades 2005; Zibechi 2006). At the same time, the pressure that the Native Community framework exerts on Indigenous peoples to adapt their livelihoods changes their necessities towards the type of development that IIRSA appears to offer (Greene 2009; Harvey and Knox 2015; Jarrier 2012; Kanai 2016; Oliart and Biffi 2010; Ráez Luna 2017; Zibechi 2006). The testimonies gathered by Eduardo in off-road communities illustrate this point, as Native Community members have the urgent need to sell agricultural produce, and presumably timber as well, in order to access cash to buy the processed foods and goods that their new lifestyles require. As mentioned before, the process of acculturation, alongside dispossession and environmental degradation, that Indigenous people have gone through in Manu fosters aspirations for the modernization that first roads promise to the frontier (Harvey and Knox 2012; Oliart and Biffi 2010; Ráez Luna 2017). These aspirations and promises align well with the modernizing discourse behind IIRSA, which makes road building to connect to the IIRSA network appear to be a necessary step towards their fulfilment (González 2012; Harvey and Knox 2012; Jarrier 2012; Kanai 2016). Thus, for the majority of people living in Manu, the possibility to connect with the Interoceanic Highway through the gradual expansion of the Manu Road signifies the end of the feelings of isolation and dispossession they have been having for decades (Gallice et al. 2017; Harvey and Knox 2012; Uribe 2017; Wilson 2004).

As frontier roads are highly likely to benefit powerful local elites the most, their economic and political power also play in favour of the popular approval of these infrastructures and the 'road as development' discourse (Harvey et al. 2017; Harvey and Knox 2015; INRENA 2008). Eduardo's accounts about the past events in Shintuya and the current situation in Puerto Shipetiari, as well as the testimonies of timber mill owners presented previously, are evidence of the way in which the Manu Road is highly likely to favour migrant loggers. The way in which members of this local elite have taken over decision-making in the Native Community of Diamante by marrying into it and using their leadership positions to formally support the road as a communal unit, described in Sect. 7.1.1 of this chapter, is a further step in this process. Situations like these show how economic elites access political power through 'inverse governance', which allows them to use the Native Community framework and other similar mechanisms in their favour (Larsen 2016). Furthermore, the high potential to profit from road construction motivates loggers and their political allies to 'conjure' the supposed benefits that the Manu Road will bring, key elements of the 'road as development' discourse, to generate popular buy-in for this project (Harvey and Knox 2012; Tsing 2005). The account presented Sect. 7.1.2 of this chapter, where SRG authorities and leaders in logging communities construct an illusory argument to favour the construction of the road, illustrates this. The creation of an 'economy of appearances', in which the benefits of development are presented as a reality even before they happen, is also clear in some of the situations described (Harvey and Knox 2012; Tsing 2000, 2005). The construction of inoperative power lines and water-provision facilities that make the benefits of road construction an almost palpable certainty are noteworthy examples. The manoeuvre that finally guarantees political benefits from the construction of apparently beneficial infrastructures is credit-taking, which is also seen in Harvey and Knox's work on the Interoceanic Highway (2015). An example of this is shown in the following comment by Eduardo's former associate Daniel, from the MAA, 'When the previous GOREMAD authority started working on the road, it was Ricardo Gómez who opposed the project. Now that he is the appointed authority in the SRG, he has become the main proponent!' In a similar attempt, the SRG manager inaugurated the newest section of the road before it was built in order to do so before his appointed mandate ended in January 2019 (John and Munro 2019). Clearly, Gómez wanted to be the one who took this project forward. The profit of taking credit for the road project became evident later, when he and other prominent proponents of the road were elected as provincial and district mayors in the following term. These positions currently grant them with power over land titling and zoning, which is likely to provide them with further benefits.

Similarly to what Harvey and Knox observe in their work (2012, 2015), other testimonies gathered during Eduardo's fieldwork show that whenever support for the road isn't achieved through these means, it is forced by social sanctions. What Adán, the merchant from Boca Manu, said makes this quite clear, 'I can't speak my mind at the meetings. The communal authorities have already decided what they want and if I say anything against the road then I get bad looks. I can't say much if everyone wants the road. I have no choice but to say the same thing they say. The road is going to come anyway; it has been decided for political reasons'. Miki, the Matsiguenka boat driver who lived in Diamante, gave me more evidence of forced support when speaking about his participation in the communal working days organized by Diamante's community council for community members themselves to build the road. 'I also participated in that *faena*', he told me, 'they took me and others by force. They said that if everyone didn't participate then the government wouldn't continue building the road and we would be stuck here. They said we had to go no matter what. There were people that wanted to do it and others that didn't but it was done anyway'. Thus, there is an indication that verbal and physical support for the road project were mandated by communities and their leaders from those that opposed or were not interested in this infrastructure.

Through these mechanisms, the economic and political elites of Manu engender popular support for road building, or the formal appearance of it, which plays out in their favour, as well as that of the objectives of IIRSA and its attempts to control territory for resource extraction (Kanai 2016; Tsing 2005; Zibechi 2006). The compatibility between the objectives of regional infrastructure expansion strategies and the 'enchantments' of roads in the frontier, thus, stands in opposition to the restrictions that conservation seems to represent for frontier dwellers (Harvey and Knox 2012; Jarrier 2012; Kohler and Brondizio 2016; Peluso and Lund 2011; Zibechi 2006). Having established this, we now move on to our brief discussion about how local reactions towards conservation and road construction amalgamate to further reinforce support for the road.

7.2.3 Territorialization for Conservation and 'Neoliberal Territorial Design'—Tipping the Balance

'I think we should exclude the whole right [southeast] side of the Alto Madre de Dios from the buffer zone of the [Manu National] Park and of [the] Amarakaeri [Communal Reserve]. That way that land could be turned into state land under unrestricted use and become productive. Communities would be able to work without having to fight anyone. Not having SERNANP meddling would be of great help for the authorities and our communities', were Ricardo Gómez's concluding remarks at our interview at the SRG. He seemed to be quite confident about the popular support this notion would have, especially once the road was built, which he also knew would eventually happen. Communities, farmers and the timber sector were all on his side. For him and other local officials, the protected areas on both sides of the proposed road were an obstacle to the type of development they wanted to promote, a vision they shared with loggers. Dámaso, one of the timber mill owners Eduardo spoke to in Salvación, supported this view, 'One of the main limitations we have are the protected areas. All of that has cut short the progress of the timber business in this province and the wealth it could have produced for our communities. Because we are in the buffer zone we are limited in our ability to work and develop. We want to change this'.

Sharing this view meant that both local authorities and loggers had Manu's conservation interests as a common adversary, an obstacle to their political and economic gain. Popular animosity towards the conservation sector given the restrictions it imposed to local livelihoods entrenched support for road building among communities in Manu (Gallice et al. 2017; Larsen 2016; Oliart and Biffi 2010; Ráez Luna 2017; Shepard et al. 2010). At the same time, the general tendency of national policy towards infrastructural expansion and resource extraction was also in their favour (Barrantes et al. 2014; Larsen 2016; Oliart and Biffi 2010). These factors came together to reinforce the 'enchantments' of the Manu Road, guaranteeing buy-in from off-road communities despite the results they were able to observe in on-road communities (Gallice et al. 2017; Harvey and Knox 2012).

Along the previous few sections, we have established that the combined effects of the confronted efforts for territorialization in Manu ultimately generate support for the construction of the Manu Road. The widespread animosity towards conservation in Manu province is particularly relevant in this effect as it further entrenches support for the road as a reaction towards the restrictions imposed by the MNP and the ACR. In the following sections, we conduct a brief analysis of the way in which local elites and their political allies use this sentiment of rejection towards nearby protected areas to reinforce the narrative of 'roads as development'. We conclude this chapter with a section that discusses how opposing voices have little resources to contest the dominant pro-road narrative.

7.3 "You Have to Break Eggs to Make a Cake!"—Analysing the 'Road as Development' Discourse in Manu

A noteworthy tool used by road proponents to reinforce the popular support for the Manu Road among local communities and the Madre de Dios regions' general public has been their use of a discourse of 'the road as development' as a resource for conjuring (McEwan 2009; Tsing 2005). The power that discourse has to define values and shape the 'truth' has been used actively by the elites of Manu province to conjure up road construction as an unquestionable necessity (McEwan 2009; Tsing 2000, 2005; Vadjunec et al. 2011). Its power is such that it has allowed these powerful agents to further entrench this notion in the imaginaries of the communities most vulnerable to its risks, and to censor those that oppose it (Harvey and Knox 2015; McEwan 2009; Tsing 2005; Uribe 2017). Specifically, the power of the discourse of 'the road as development' has been directed against SERNANP and conservation NGOs working in Manu, to further weaken their leverage with local communities (McEwan 2009; Oliart and Biffi 2010; Vadjunec et al. 2011).

What follows is an analysis of how the political and extractive elite of Manu has used this 'road as development' discourse in their favour and how it has been contested. First, we analyse the way then regional governor Luis Otsuka conveyed the discourse of 'the road as development' along his administration,[9] targeting the conservation sector as the enemy of this model and the people of the region. In the next section, we examine the way in which provincial and local authorities echo this narrative in an attempt to portray the Manu Road as environmentally harmless in order to invalidate the arguments upon which the conservation sector opposed its construction. We conclude this discussion in the last section with an analysis of contesting voices at various scales and the effects of the lack of available outlets that they face.

7.3.1 "You Don't Live like that!"—How the Discourse of 'Conservation as Colonization' Is Used to Support the 'Road as Development' Discourse

At the end of 2014, a regional electoral year, the Law Proposal that sought to declare the Manu Road a 'public necessity' was turned into a battle cry by Luis Otsuka's campaign, as was his opposition to conservation. As is usual from the biggest cities in the country to the smallest district capitals, the streets of Salvación were overcome by a cacophony of *cumbia* songs with modified lyrics that supported each candidate, which were blasted from speakers within improvised campaign offices. In between each song played from Otsuka's campaign locale, a fraction of one of his speeches would be heard: 'Why is the ministry of the environment giving our land and our resources to [these foreigners][10]? Why are we being prevented from using our own resources to overcome our poverty? We are tired of this! My government will strive to provide our people with access to our resources! We will build the roads you have been demanding for so long! We will start caring more for our people than for the beasts of the forest that these foreigners care for!' Building roads to connect the region's districts to the network that stemmed from the Interoceanic Highway became one of his main objectives, and his outspoken opposition to conservation organizations, governmental or non-governmental, became his most aggressive rhetoric. As a leader of the region's mining sector, his main proposition to boost development in the region was based on resource extraction, which explained his position on roads and conservation (Torres López 2016; Walker 2012).

---

[9] 2015–2018, this period encompasses most of the issues discussed in this case study.

[10] A specific NGO with large conservation concessions in Manu province and around the MNP was mentioned.

Once elected, Otsuka and his government continued relaying this kind of discourse, which portrayed the conservation sector of the region as a fatal enemy of development and, thus, of Indigenous peoples. 'We are doing this to help the native communities. Would you prefer that they die, that they rot here! (sic.)', was his answer when journalist Fabiola Torres López asked about the 'real beneficiaries' of the Manu Road (2016). He frequently used this narrative, in the most graphic way possible, whenever his haste to build roads throughout the region was questioned or officially supervised. 'We need to take these [GOREMAD] rulings forward to make development viable through our natural resources. A child cannot live happily when they're crying of hunger, when they're vomiting blood with tuberculosis. Is that how you live? You don't live like that! I can tell you because I went to school naked! That is why I want rules that recognize us so one day we can live dignified lives as Peruvians with our resources', was how he addressed members of the national congress at a supervisory meeting (Telepuerto Noticias 2017). He continued by alluding to his involvement with the Manu Road, 'I'm facing trial now because 14 NGOs[11] demanded this from president Humala. I am facing trial for opening a road to the poorest communities, those that have nothing to eat and face a high rate of malnutrition. Who do you think is behind all this? The NGOs, the most powerful ones that come from other latitudes and only look for their own profit, not for the benefit of us Peruvians!' (Telepuerto Noticias 2017).

Descriptive narratives of this sort were also broadcast through his government's media outlets and conveyed through regional policy documents. The blame for the region's poverty was placed squarely on the conservation sector, public and private. An example of this is a video posted onto GOREMAD's social media in 2017 in which images of the Matsiguenka communities inside the MNP were used in a rather exploitative manner (GOREMAD 2017). 'A reality that we cannot hide. The Indigenous families of Manu province are immersed in extreme poverty. The prohibitions of SERNANP, the elusive hands of NGOs and the idleness of the state condemn thousands of Amazonian native families in Madre de Dios to misery', was what the subtitles to these images said, accompanied by a dramatic score (GOREMAD 2017). The subtitles continued by saying that governor Otsuka was 'committed to fight against greatly powerful interests to achieve the quality of life that our native brothers aspire to' (GOREMAD 2017). This narrative later became part of regional policy guidelines published early in Otsuka's last year in office (GOREMAD 2018b). The 'Pact for the Economic and Social Development of Madre de Dios'[12] mentioned the region's abundance of untapped natural resources and, ironically, its rich biodiversity (GOREMAD 2018b). Despite mentioning biodiversity as a boon, it also sought to avoid 'manipulation from the environmental sector', namely MINAM and NGOs in general, and accused them of receiving 'foreign funding' in an attempt to portray them as foreign abusers (GOREMAD 2018a, b). It is evident, then, that Otsuka's

---

[11]Those who signed the letter in opposition to the Manu Road project, mentioned in Sect. 7.1.2 of this chapter.

[12]*Pacto por el Desarrollo Económico-Social Madre de Dios* (GOREMAD 2018b).

GOREMAD, which directly guided and funded the actions of the SRG, emphatically blamed the conservation sector for the conditions that off-road communities in Manu and the rest of the region faced.

The condemnation of this sector clearly serves the interests of the economic and political elites that promote road building in Manu. The bluntness of the discourse used by different government instances in Madre de Dios against the conservation sector and its portrayal as a group of foreign agencies that restrict local livelihoods go far back (Oliart and Biffi 2010). This narrative reinforces the image of conservation as a colonializing interest that seeks to use local resources for the benefit of powerful foreign 'landlords' (Adams and Hutton 2007; Grosfoguel 2007; Peluso and Lund 2011; Robbins 2004). Thus, the animosity generated by the asymmetrical relations between local communities and this sector, especially regarding the establishment of protected areas like the MNP, is easily exploited by the extractive elites that influence and form part of the region's governmental hierarchy (Larsen 2016; Oliart and Biffi 2010; Peluso and Lund 2011; Ráez Luna 2017; Salisbury et al. 2011). This feeds the conflict generated by the restrictions on which the conservation of the MNP and other protected areas are based upon, putting modernizing interests on the side of local communities (Larsen 2016; Ráez Luna 2017; Robbins 2004; Shepard et al. 2010). This is evident in several of the testimonies provided in this work, which present SERNANP and the organizations that oppose road construction as an enemy of local communities, proving the success of the message transmitted by authorities like Otsuka. Furthermore, Otsuka's GOREMAD bolsters its message through the portrayal of the Indigenous communities of Manu as a subaltern 'others' in crisis (McEwan 2009; Mignolo 2000). This justifies modernizing interventions like road building and resource exploitation as the means by which more 'civilized' sectors of Peruvian society solve the crisis of Indigenous communities (McEwan 2009; Morel 2014). The distance between proponents of conservation and local communities, intensified by the restrictions established by the former, becomes quite relevant here (Larsen 2016; Oliart and Biffi 2010; Ráez Luna 2017; Shepard et al. 2010). This distance generates the perception of the Madre de Dios local and regional governments as entities situated politically, physically and ethnically 'closer' to Indigenous Amazonian communities, which legitimizes their interventions (Ames 2010; Morel 2014). This is another example of how the restrictive and exclusionary conservation strategies applied by the conservation sector in Manu bounce back against it (Gallice et al. 2017; Kohler and Brondizio 2016; Larsen 2016; Oliart and Biffi 2010; Ráez Luna 2017; Shepard et al. 2010). This facilitates the continued 'colonization', in the sense of the 'Conquest of Peru by Peruvians', of Indigenous territories in the area by extractive elites through road expansion, as well as the social and environmental consequences that this carries (Gallice et al. 2017; Larrea-Gallegos et al. 2016; Larsen 2016; Morel 2014).

In this section, we have established the way in which the GOREMAD authorities at the time used the image of conservation as a colonizing force to support the 'road as development' discourse. Having done this, we now move on to exploring the tactics through which local authorities and other road proponents conveyed this discourse by attempting to undermine the arguments used by the conservation sector to oppose the road.

7.3.2 *Environmental Harmlessness—The Discourse of Local Leaders*

The narrative relayed by Otsuka's GOREMAD is supported locally through other, perhaps less aggressive, means. Representatives of the Manu SRG and other local authorities often argue for the environmental harmlessness of the Manu Road through different claims. The very likely environmental consequences of the construction of this infrastructure show that such claims may constitute further attempts at 'conjuring' benefits that the road is unlikely to generate (Gallice et al. 2017; Larrea-Gallegos et al. 2016; MINAM 2016; Tsing 2005). Additionally, the apparent benefit of an environmentally harmless road further delegitimizes SERNANP's and the conservation sector's arguments against road construction in the area.

Several authorities and community leaders directly denied any possibility of environmental impact to the MNP and ACR given that the road would not pass through them. Duncan, the young council member from Diamante, said, 'There won't be any damage done to the Park from the road because it will go through our land. Why would they oppose the road if it wasn't going to go through their area?'. This notion originates in the 2013 Law Proposal, which states that the Manu Road, 'Does not go through the Reserved Natural Zone, the National Park or their buffer zones' (Congreso de la República 2013, p. 5). The first two assertions are true, the road does not go through the core area of the ACR or of the MNP but, contrary to this statement, it does go through the buffer zones of both protected areas (see Map 2.1) (Finer et al. 2016; Gallice et al. 2017; Humala and Cateriano 2015; MINAM 2016). The issue may be that communal territories and the buffer zones of these protected areas overlap, which is another example of the restricted control that Indigenous peoples have over the territories they are granted by the State (Larsen 2016; Oliart and Biffi 2010).

Other authorities portrayed the economic activities the road would promote as perfectly sustainable. Aníbal, the deputy mayor of Fitzcarrald, told Eduardo about the timber 'capture' method used in Boca Manu. Members of this community had formed an association that was allowed to bring in the fallen trees that floated down the Manu river and use them as timber instead of cutting down live trees. Eduardo had witnessed how this was done in previous occasions. He writes:

> On the occasions I had visited Boca Manu before, I had seen boat drivers placing their vessels next to the enormous trees that floated down the river, carefully tying them to their boats and bringing them in. It seemed to me like fluvial whale hunting. "This method will allow us to continue working with timber without causing more deforestation. The road will allow us to make a better profit from that!", was district deputy mayor Aníbal's conclusion. Silvia, the merchant, later told me that, "Not everyone here is a member of that association. Only the few people who signed up for that originally are authorized to do that. If others try to do it

then the associates will not let them. Everyone else cuts trees without a permit. How do you think all the timber mills here would operate with just the stuff they bring in from the river?"

It seemed the deputy mayor was using the example of this association to reduce the perception of the deforestation that the road would potentially cause. However, Aníbal's argument was only mentioned by him. As seen in the account presented in Sect. 7.1.2 of this chapter, other allegations about the environmental harmlessness of the economic activities the Manu Road would facilitate were echoed by several people in leadership positions. There seemed to be a concerted effort, mostly by Andean migrants in leadership positions, to support the argument about how the road would make farming profitable and motivate loggers to switch to this business. The community council in Diamante, SRG manager Ricardo Gómez and *Profesor* Paniagua, among other people in leadership positions, had shared this logic in very similar terms. Gómez and Mr. Castro, the President of Nuevo Edén, had gone far to make this argument appear true by organizing an agricultural fair in a community that, according to Ricky and Nidia, would hardly make the effort to harvest their bananas for sale. Eduardo's account shows how other informants further evidenced the flawed logic of this argument. The fact that the road would be an asset for the logging sector in Manu, discussed in Chap. 6, is also evidence against this well-coordinated argument. What is more, it is important to keep in mind that this argument considered farming as an environmentally friendly activity, at least friendlier than logging. However, both activities are environmentally damaging consequences of increased road access, which meant that damage to the ecosystem would happen either way (Carvalho et al. 2002; Gallice et al. 2017; Larrea-Gallegos et al. 2016; Soares-Filho et al. 2006 cited in Vergara et al. 2014).

Further attempts at reducing the perception of the environmental damage the Manu Road is likely to cause were focused on the terms used to refer to it (Gallice et al. 2017; Larrea-Gallegos et al. 2016). Officials from the SRG and other institutions did not call it a 'road',[13] as most ordinary people did, but called it a 'local track',[14] and insistently encouraged others to do so. At the council meeting Eduardo witnessed in Diamante, Chief Carmen conveyed a message from one of the SRG officials appointed to the road project, whom she and Neptalí had met at the meeting they attended to on their own. '*Economista* Francesca says that we should call it a "local track", not a "road". A "road" is what was built up to Nuevo Edén, ours is just a "local track"', she said to her council members. They must have been able to see that there is no visible difference between the unpaved track before and after Nuevo Edén (Larrea-Gallegos et al. 2016). The term 'local track' is commonly used to describe small dirt tracks built to provide vehicle access to farming plots and bring products out to larger roads or river ports. According to the terminology used by the Peruvian Ministry of Transport (MTC), it is the lesser denomination of transport infrastructures fit for motor-vehicles (Díaz and Miranda 2014; El Peruano 2007). The gradual changes in the terms used along the progress of the Manu Road controversy are interesting

[13] *Carretera.*

[14] *Camino Vecinal.*

to consider. The 2013 Law Proposal calls it a 'regional integration road',[15] which fits MTC's second highest denomination of transport infrastructures (Congreso de la República 2013, p. 1; Díaz and Miranda 2014). Other official documents call it a 'penetration road',[16] a term commonly used to denominate the frontier roads built for the 'colonization' of the Peruvian Amazon in the twentieth century, which reveals its original purpose as a road that would facilitate extractive activities (Harvey and Knox 2012; INRENA 2008; Morel 2014). Thus, the changes in the terms used to describe the Manu Road, which so far is wide enough for large logging trailers to use, reveal a push to euphemize the vocabulary used to refer to it (Larrea-Gallegos et al. 2016). This responds to the larger objective of portraying this road as less environmentally harmful than it will actually be (Gallice et al. 2017; Larrea-Gallegos et al. 2016).

Additionally, despite efforts to make the consequences of the road seem harmless to the ecosystem and the contrasting situation seen on the ground and confirmed by the literature, the road project did not contemplate any proactive effort to curb its environmental effects. The road project itself needed to include actions to control the environmental impact of the road but Ricardo Gómez from the SRG left this to be determined entirely by the National Agricultural Sanitation Agency,[17] in charge of evaluating the project. 'Only if they tell us to conduct a detailed assessment we will have to include an environmental impact assessment', he said. As mentioned previously, entities in charge of enforcing environmental law were either absent or under-resourced in Manu, which left controlling the road's environmental effects in the passive hands of the SRG. Considering that the capacity to enforce environmental law is key to conservation, the uncontrolled effects of the road are likely to generate significant environmental damage that will not be controlled (Arima et al. 2013; Carvalho et al. 2002; Gallice et al. 2017; Larrea-Gallegos et al. 2016). Rather than that, the environmental damage that the road is likely to cause was revealed to be considered a worthy cost of development by more open informants, something that is common in similar cases (Arima et al. 2005; Tsing 2009). Pedro, the timber mill owner, gave some insight into this point of view quite eloquently: 'I think the road will cause an environmental impact. That's for sure! But that's the cost of development. Don't people say that you have to break eggs to make a cake?"

The significant environmental impact that the extension of the Manu Road would cause is as clear in the literature as it was to the users of the existing road, who could see firsthand the way in which loggers operated and the amount of timber being driven through it (Gallice et al. 2017; Larrea-Gallegos et al. 2016). However, it was considered to be a cost of development that the communities of Manu had to incur in. Thus, the opposition from the conservation sector for environmental reasons, as well as the organizations that were part of this sector, had to be discredited as much as possible in order to legitimize the 'road as development' discourse. As the accounts Eduardo gathered show, this was the object of an organized effort between the authorities and powerful elites of the province and the Madre de Dios region.

---

[15] *Carretera de Integración Regional.*

[16] *Carretera de Penetración.*

[17] *Servicio Nacional de Sanidad Agraria (SENASA).*

In this section, we have analysed the ways in which local authorities, leaders and the logging elite have attempted to undermine the arguments upon which conservation based its opposition to the road project in order to support the 'road as development' discourse. In the following section, we explore the voices that opposed this discourse and the apparent lack of instruments through which to convey their message of contestation.

7.3.3 Silenced Opposition—Contesting Arguments and Their Lack of Local Leverage

The presence of a dominant discourse of modernity, such as that of 'the road as development', often prompts the emergence of alternative views that may be capable of contesting it (Escobar 2008; McEwan 2009). The views of those who resist the negative impact of imposed development often contribute to the formation of alternative solutions to dealing with the issues that development strategies attempt to address (Cupples et al. 2019), like the lack of economic opportunities and public services Eduardo observed in Manu. However, the silencing of these voices, based on the entrenchment of certain notions of development and the binaries of modernization, is also a characteristic element in these debates, even more so when these contesting voices are of Indigenous origin (Aikau and Spencer 2007; Grosfoguel 2007; Gudynas 2016; McEwan 2009). In the case of Manu, a lack of effective outlets has prevented these voices from successfully resisting the modernizing discourse of 'the road as development' that is conveyed by the extractive elites of Manu and Madre de Dios.

Our observations show that local voices that contest the expansion of the Manu Road in its current form have in fact emerged but only to be excluded from the local debate. The constant reminders from Soledad and others from Shintuya about the loss of their culture and resources because of road expansion is an example of such voices. The dismissal of these risks from people in off-road communities, like Duncan from Diamante who assured communities would be able to resist land encroachment, shows the capacity of the discourse of 'the road as development' to overpower contesting arguments like these. Moreover, the case of Adán from Boca Manu, who feared social sanctions if he spoke publicly of his fears regarding the road, reveals the active silencing of contesting arguments, apart from passive dismissal. Additionally, the case of Miki from Diamante, who was forced to work on road construction despite opposing it, shows how support for the road is actively mandated by community elites. These examples illustrate how opposing voices and actions from local agents are either ignored or actively silenced by road proponents. This situation is further evidence about how the political power held by proponents of the Manu Road facilitated the circumvention of effective participative measures in the process of road construction, which has been pointed out by the Native Federation of the Madre de Dios River and Effluents[18] (FENAMAD) (Gallice et al. 2017). This

[18] *Federación Nativa del Río Madre de Dios y Afluentes.*

type of avoidance of local consultation is also part of the strategy used to implement the Interoceanic Highway and other IIRSA projects (Harvey and Knox 2015; Zibechi 2006), the overall intention of which is reproduced by proponents of the Manu Road project. Thus, the exclusion of contesting voices like these prevented alternatives to the outright opposition to road construction proposed by the conservation sector from being considered in the local debate (Gallice et al. 2017; MINAM 2016).

Examining the role of multi-ethnic Indigenous organizations as vehicles for these contesting voices is another alternative to gauge their influence, or lack thereof, in the debate around the Manu Road. It is important to consider that the heterogeneity, and even contradiction, of messages between different levels of Indigenous organizations is a characteristic of the Peruvian and Amazonian Indigenous movement (Larsen 2016; McNeish 2013; Oliart and Biffi 2010). This heterogeneity in perspectives and allegiances is also one of the reasons why the authority of Indigenous federations is not universally respected (Larsen 2016). An example of this is the participation of FENAMAD, an organization that officially represents the Indigenous peoples of the Madre de Dios region, in local meetings about the road project. In them, the organization's representatives recognized the benefits of road connectivity but also protested the lack of consultation during the construction of the stretch built in 2015 and warned about how the road had mostly been a liability for Native Communities so far (Gallice et al. 2017; MINCUL 2016a). Despite their attempt to open the discussion for alternatives, minutes from these meetings show that FENAMAD's interventions in this regard were mostly overpowered by voices that supported the road, mainly from community Chiefs and other relatively powerful agents (MINCUL 2016a, b, c). The ability of local road proponents to silence the representatives of a regional Indigenous federation illustrates how effective the 'participative' aspects of these meetings were. In debates at the national level, the message of AIDESEP, which is supposed to represent the interests of Indigenous organizations nationally, focused on how risky first road projects would be for isolated Indigenous peoples who could contract new illnesses through increased contact (AIDESEP 2018a, b). Having caused deaths and becoming a constant threat for Shipetiari and Diamante, isolated groups were seen by most Native Communities in Manu as bands of 'savage' raiders,[19] which made their safety an unlikely element in arguments put forth by local voices (García Delgado 2018; Torres López 2016). AIDESEP's focus on the safety of isolated Indigenous peoples, a more vulnerable population that is appealing nationally and globally but that is seen negatively in Manu, reveals much about the lack of influence that local voices have in the larger scale debate (AIDESEP 2018a, b). Thus, local voices that proposed alternatives to what was being put forth by road proponents or the conservation sector didn't have a chance to be heard through the organizations that were supposed to represent them.

Local voices that opposed road construction were not incorporated by representatives of the conservation sector into the debate either, despite their similar interests.

[19] A few of the informants agreed that the 'calatos', or naked-people, as isolated Indigenous people are colloquially referred to in Manu, should be 'civilized' through forced contact to avoid further violent encounters.

The National Government and the MINAM, during the mandate that took legal action against the construction of the road, produced several documents that explained their opposing arguments but did not include local voices or address the needs of local communities (Humala and Cateriano 2015; MINAM 2016; SERNANP 2017). Their opposition was based on purely environmentalist arguments, like the protection of the constitutional right of Peruvians to benefit from the preservation of the natural environment (Humala and Cateriano 2015; MINAM 2016). According to the National Government in turn, this right took priority over economic interests, which were mostly based on illegal activities (Humala and Cateriano 2015; MINAM 2016). In this case, the economic interests alluded to were not those of large extractive companies, the usual antagonist of the conservation sector, but those of local communities and extractive interests seen as closer to, and even representative of, communities. This, again, established the position of the conservation sector and the entities of the National Government supporting it as powerful foreign forces that work in favour of extra-local interests (Adams and Hutton 2007; Grosfoguel 2007; Peluso and Lund 2011; Robbins 2004). It is important to point out that none of the official documents opposing the road project proposed solutions to the conditions local communities faced, except for mentioning the effects of the road expansion on the tourism industry, which has very marginal effects on said conditions (MINAM 2016). No alternatives were provided to address the issues that the lack of road connectivity generated for local communities or to reduce the environmental impact of the road once it was built. Alternatives of this nature, like local subsidies for riverboat fuel and the intensification of agriculture in already established agricultural land, mentioned by Gallice et al. (2017), were not contemplated at all. Thus, the message of MINAM and the National Government of the time, in the representation of the conservation sector, remained distant from the needs of local communities. By excluding local voices that opposed road construction, these government entities outright opposed road construction lacking to propose alternatives to the Manu Road that seemed viable to local eyes.

The examples we have provided in this section show that several voices that contested the 'road as development' discourse of Manu's extractive elites did emerge locally. Nonetheless, these voices were either ignored or silenced in local fora, given the political power of road proponents and the entrenchment of their modernizing narrative (Grosfoguel 2007; Gudynas 2016; Harvey and Knox 2015; McEwan 2009). Additionally, the message of these voices was mostly ignored by multi-ethnic Indigenous organizations and government entities in charge of conservation, who chose to focus their opposition to the road on arguments not aligned with local interests. So, what most of these organizations expressed remained as alternatives proposed by distant agents with little interest on, or understanding of, the communities of Manu and their livelihoods. These remained aligned to the restrictive preservation of MNP as the 'crown jewel' of the Peruvian conservation sector, thus, having little chance of being taken up locally (Gallice et al. 2017; Shepard et al. 2010). In this way, the lack of effective incorporation of local voices that opposed road construction kept local interests separate from alternatives to the 'road as development' narrative, giving way to adamant local support for it (Gallice et al. 2017).

Therefore, this final section has demonstrated how local voices opposing the 'road as development' discourse have had little effective means to convey their message. This has resulted in a situation in which the pro-road discourse has been adopted by most communities and local leaders, generating enough local support for road construction to become an overwhelming force in the province.

This chapter has explained how the confrontation to territorialize the land that formerly sustained the lives of Manu's Indigenous peoples between the conservation sector and the forces behind resource extraction has tilted in favour of the latter. The conservation sector has restricted local livelihoods in Manu, in their former and current variations, by territorializing the land through the establishment of protected areas. This has perpetuated the image of conservation as a colonizing force that imposes the will of foreign interest and stifles local communities (Barrantes and Glave 2014; Díaz and Miranda 2014; Oliart and Biffi 2010; Ráez Luna 2017; Shepard et al. 2010). It's lack of consistency with the will of Indigenous peoples, other local residents and the message of their organizations has only worsened this situation (Oliart and Biffi 2010; Ráez Luna 2017; Shepard et al. 2010). On the other hand, the territorialization promoted by logging elites and authorities allied to them through road building influenced by IIRSA is likely to cause conflict and deprivation, especially for Indigenous communities (Harvey and Knox 2015; Kanai 2016; Pieck 2013; Tsing 2000, 2005; Zhouri 2010; Zibechi 2006). Regardless of this, these actors have been able to conjure an image of modernity and economic prosperity among local people through the discourse of 'the road as development'. This has reinforced the 'enchantments' of the Manu Road and further entrenched its characterization as the only solution to the area's lack of development. Local voices that oppose the expansion of the Manu road are also present but are swiftly and forcefully shut out of the debate. Having demonstrated this, we now move on to our concluding chapter.

References

Adams WM, Hutton J (2007) People, parks and poverty: political ecology and biodiversity conservation. Conserv Soc 5(2):147–183

AIDESEP (2018a) Informe sobre la situación de los pueblos en aislamiento y contacto inicial de la Amazonía peruana. Lima, AIDESEP

AIDESEP (2018b) Pronunciamiento de AIDESEP sobre la LEY N° 30723, http://www.aidesep.org.pe/pronunciamiento-de-aidesep-sobre-la-ley-no-30723/

Aikau HK, Spencer JH (2007) Introduction: local reaction to global integration-the political economy of development in indigenous communities. Altern Global Local Polit 32(1), 1–8

Ames P (2010) Desigualdad y Territorio en el Perú: una geografía jerarquizada. Argumentos 1

Arima EY, Walker RT, Perz SG, Caldas M, Arima EY, Walker RT, Caldas M (2005) Loggers and forest fragmentation: behavioral models of road building in the Amazon Basin. Ann Assoc Am Geogr 95(3):525–541

Arima EY, Walker RT, Souza C, Pereira R, do Canto O (2013) Spontaneous colonization and forest fragmentation in the Central Amazon Basin. Ann Assoc Am Geogr 103(6):1485–1501

Baletti B (2012) Ordenamento territorial: neo-developmentalism and the struggle for territory in the lower Brazilian Amazon. J Peasant Stud ISSN 39(2):573–598

Barrantes R, Fiestas J, Hopkins Á (2014) Evolución de la infraestructura de Transporte y Energía en la Amazonía Peruana. In: Barrantes R, Glave M (eds) Amazonía Peruana y Desarrollo Económico. GRADE, IEP, Lima, pp 109–160

Barrantes R, Glave M (2014) Introducción. In: Barrantes R, Glave M (eds) Amazonía Peruana y Desarrollo Económico. GRADE, IEP, Lima, pp 13–20

Bluwstein J, Lund JF (2018) Territoriality by conservation in the Selous-Niassa Corridor in Tanzania. World Dev 101:453–465

Carvalho GO, Nepstad D, McGrath D, Diaz MDV, Santilli M, Barros AC (2002) Frontier expansion in the Amazon: balancing development and sustainability. Environ Sci Policy Sustain Dev 44(3), 34–44

Congreso de la República (2013) Proyecto de ley que Declara de Necesidad Pública e Interés Nacional la Construcción de la Carretera de Integración Regional de Madre de Dios. Comisión de Transportes y Comunicaciones, Lima

Cupples J, Prieto M, Palomino-Schalscha M (2019) Latin American development-editor's introduction. The Routledge handbook of Latin American development. Routledge, London, pp 1–11

Díaz R, Miranda J (2014) Áreas Naturales Protegidas en el Perú: Efectos sobre la Deforestación y su Relación con el Bienestar de la Población Amazónica. In: Barrantes R, Glave M (eds) Amazonía Peruana y Desarrollo Económico. GRADE, IEP, Lima, pp 209–245

El Peruano (2007) Aprueban Reglamento de Jerarquización Vial, https://portal.mtc.gob.pe/transportes/caminos/normas_carreteras/MTCNORMAS/ARCH_PDF/Regl.deJerarquizaciónVial.pdf

Escobar A (2008) Territories of difference: place, movements, life, redes. Duke University Press, Durham and London

Finer M, Novoa S, Olexy T (2016) Construction of a new highway between Manu National Park and Amarakaeri Communal Reserve (Madre de Dios), maaproject.org

Gallice G, Larrea-Gallegos G, Vásquez-Rowe I (2017) The threat of road expansion in the Peruvian Amazon. Oryx, 1–9

García Delgado F (2018) Comunero muere por impacto de flechas de indígenas mashco piro. El Comercio Website, https://elcomercio.pe/peru/madre-de-dios/contactados-matan-flechazos-comunero-peru-noticia-501491

González GL (2012) El derecho ambiental en la integración regional de Suramérica. Letras Verdes 12:42–56

GOREMAD (2017) Video on Poverty of MNP Indigenous Communities, https://www.facebook.com/GOREMAD.MDD/. Accessed 12 Dec 2017

GOREMAD (2018a) Manifiesto de Madre de Dios: Pacto por el desarrolllo económico-social y por un Madre de Dios sin Pobreza ni Discriminación. Gobierno Regional de Madre de Dios, Puerto Maldonado, Madre de Dios

GOREMAD (2018b) Pacto por el Desarrollo Económico-Social y por un Madre de Dios sin Pobreza ni Discriminación. Puerto Maldonado, Madre de Dios

Greene S (2009) Caminos y Carretera: Acostumbrando la indigenidad en la selva peruana. IEP Instituto de Estudios Peruanos, Lima

Grosfoguel R (2007) The Epistemic Decolonial Turn. Cultural Studies 21(2–3):211–223

Gudynas E (2016) Beyond varieties of development: disputes and alternatives. Third World Q 37(4):721–732

Harvey P, Jensen CB, Morita A (2017) Introduction: infrastructural complications. Infrastructures and social complexity: a companion. Routledge, London, pp 1–42

Harvey P, Knox H (2012) The enchantments of infrastructure. Mobilities 7(4):521–536

Harvey P, Knox H (2015) Roads: an anthropology of infrastructure and expertise. Expertise, Ithaca, N.Y

Humala O, Cateriano P (2015) Oficio 243-2015-PR. Congreso de la República, Lima

INRENA (2008) Plan Maestro de la Reserva Comunal Amarakaeri 2008–2012. Ministerio de Agricultura, Lima

Jarrier A (2012) Seminario Internacional: "Todos los caminos conducen al Brasil. La geopolítica del plan de Integración de Infraestructura Regional Suramericana-IIRSA-en la región andina." In Bulletin de l'Institut français d'études andines, vol 41, pp 303–307
John B, Munro E (2019) Voices on an Amazon road-report to funders
Kanai JM (2016) The pervasiveness of neoliberal territorial design: Cross-border infrastructure planning in South America since the introduction of IIRSA. Geoforum 69:160–170
Kohler F, Brondizio ES (2016) Considering the needs of indigenous and local populations in conservation programs. Conserv Biol 31(2):245–251
Larrea-Gallegos G, Vazquez-Rowe I, Gallice G (2016) Life cycle assessment of the construction of an unpaved road in an undisturbed tropical rainforest area in the vicinity of Manu National Park, Peru. Int J Life Cycle Assess 11(17/2016):1–16
Larsen PB (2016) Derechos indígenas, gobernanza ambiental y recursos en la Amazonía peruana, Hacia una antropología de la posfrontera. Instituto de Estudios Peruanos, Lima
Leal DB, Salisbury DS, Silva J (2015) Ideas cambiantes sobre territorio, recursos y redes politicas en la Amazonía indigena: un estudio de caso sobre Perú. J Latin Am Geogr 14(2):181–204
McEwan C (2009) Postcolonialism and development. Routledge, London
McNeish J-A (2013) Extraction, protest and indigeneity in Bolivia: the TIPNIS effect. Latin Am Caribb Ethn Stud 8(2):221–242
Mignolo WD (2000) Local histories/global designs: coloniality, subaltern knowledges, and border thinking. Princeton University Press, Princeton and Oxford
MINAM (2016) Dossier: ¿Por qué no debe aprobarse el predictamen de insistencia de la ley sobre la carretera Madre de Dios, tramo Nuevo Edén-Boca Manu-Boca Colorado?. Ministerio del Ambiente, Lima
MINCUL (2016a) Minutas 3 Reunión Carretera 17 05 16. Not published
MINCUL (2016b) Minutas 4 Reunión Carretera 21 07 16. Not published
MINCUL (2016c) Minutas 5 Reunión Carretera 04 08 16. Not published
Morel J (2014) De una a muchas Amazonías: Los Discursos sobre 'La Selva' (1963-2012). In: Barrantes R, Glave M (eds) Amazonía Peruana y Desarrollo Económico. GRADE, IEP, Lima, pp 21–46
Oliart P, Biffi V (2010) Territorialidad indígena, conservación y desarrollo: Discursos sobre la biodiversidad en la Amazonía peruana. Instituto del Bien Común, Lima
Peluso DM, Alexiades MN (2005) Indigenous Urbanization and Amazonia's post-traditional environmental economy. Tradit Dwell Settl Rev 16(2):7–16
Peluso NL, Lund C (2011) New frontiers of land control: introduction. J Peasant Stud 38(4):667–681
Pieck SK (2013) Asphalt dreams: road construction and environmental citizenship in Peru. Dev Chang 44(5):1039–1063
Ráez Luna E (2017) Pueblos Indígenas y Conservación de la Naturaleza: ¿Debates teóricos o urgencias vitales? El caso del Parque Nacional del Manu. In: Ráez Luna E (ed) Derechos Ancestrales y Conservación de la Naturaleza en Debate: El caso de los Pueblos Indígenas del Parque Nacional del Manu. Seminario Permanente de Investigación Agraria (SEPIA), Lima
Raycraft J (2018) Circumscribing communities: marine conservation and territorialization in southeastern Tanzania. Geoforum 100(2019):128–143
Robbins P (2004) Political ecology: a critical introduction. Blackwell Publishing, Oxford
Salisbury DS, López JB, Vela Alvarado JW (2011) Transboundary political ecology in Amazonia: history, culture, and conflicts of the borderland Asháninka. J Cult Geogr 28(1):147–177
SERNANP (2017) Oficio N°620–2017-SERNANP-DGANP. Ministerio del Ambiente, Lima
Shepard GH, Rummenhoeller K, Ohl-Schacherer J, Yu DW (2010) Trouble in Paradise: Indigenous populations, anthropological policies, and biodiversity conservation in Manu National Park, Peru. J Sustain For 29(2–4):252–301
Telepuerto Noticias (2017) GOBERNADOR REGIONAL SE RETIRO DE ASAMBLEA DE FISCALIZACIÓN, https://www.facebook.com/telepuerto.mdd/videos/906685402819866/
Torres López F (2016) The highway tearing through the heart of the Amazon. Ojo Público, 1–22, http://ojo-publico.com/sites/apps/amarakaeri-parte1-english/

Tsing AL (2000) Inside the economy of appearances. Public Cult 12(1):115–144
Tsing AL (2005) Friction: an ethnography of global connection. Princeton University Press, New Jersey
Tsing AL (2009) How to make resources in order to destroy them (and then save them?) on the salvage frontier. In: Histories of the future, pp 51–73
Uribe S (2017) Frontier road: power, history and the everyday state in the Colombian Amazon. Wiley, New Jersey
Vadjunec JM, Schmink M, Greiner AL, Vadjunec JM, Schmink M, New ALG (2011) New Amazonian geographies: emerging identities and landscapes. J Cult Geogr 3631(1):1–20
Vergara K, Figallo M, Glave M (2014) Infraestructura en la Amazonía Peruana: una propuesta para proyectar cambios en la cobertura boscosa en la carretera Pucallpa-Cruzeiro do Sul. In: Barrantes R, Glave M (eds) Amazonía Peruana y Desarrollo Económico. GRADE, IEP, Lima, pp 161–208
Walker R (2012) Peru's gold rush pits illegal miners against government, https://www.bbc.com/news/world-latin-america-18524330
Wilson F (2004) Towards a political economy of roads: experiences from Peru. Dev Chang 35(3):525–546
Zhouri A (2010) "Adverse Forces" in the Brazilian Amazon: developmentalism versus environmentalism and Indigenous rights. J Environ Dev 19(3):252–273
Zibechi R (2006) IIRSA: la integración a la medida de los mercados. Ecología Política 31:19–25

Chapter 8
Conclusions

Abstract Based on the analysis developed in the previous chapters, which weaves relevant literature with extensive empirical material, this concluding chapter brings to the fore the key elements that strengthen the powers of the Manu Road to 'enchant'. In particular, we focus on the implications of its location near protected areas, which are seen locally as restrictive impositions by foreign interests and mobilized by local elites as arguments in favour of a road expansion project that will mostly be for their own benefit.

Keywords Manu road · First roads · Frontier infrastructure · Enchantments · Indigenous · Amazonia · Political ecology · Conservation · Territoriality · Development

The exploration of the tensions around the Manu Road developed throughout this book has shown the aspirations, disjunctures, colonial legacies and crucial absences that shape its current state. Grounded on rich empirical material and an extensive revision of the context shaping this case study, we have not only demonstrated the 'enchantments of infrastructure' (Harvey and Knox 2012) at work here, but also extended, complicated and complemented this concept with other considerations. In particular, we have highlighted how the unpaved tracks that connect the frontier to large road networks may become even more 'enchanting' when they are located near protected areas. Even when their main focus is to expand the reach of extraction, the emergence of regional infrastructural networks adds to the apparent prosperity that road connectivity is believed to be able to generate and provides it with territorializing force (Kanai 2016; Zhouri 2010; Zibechi 2006). As an extension of these networks, first roads seem to offer frontier communities the chance to break away from the various restrictions that protected areas have introduced in their livelihoods (Adams and Hutton 2007; Kohler and Brondizio 2016; Oliart and Biffi 2010; Ráez Luna 2017; Shepard et al. 2010). Furthermore, the restrictive nature of protected areas, and their image as an imposition of foreign interests, is exploited by local extractive elites that are most likely to reap the highest benefits from road connectivity through the 'road as development' discourse (Adams and Hutton 2007; Harvey et al. 2017; Harvey and Knox 2015; Kohler and Brondizio 2016). These elements strengthen the aspirations that give way to the 'enchantments' of roads, increasing their capacity to appear

E. Salazar Moreira and M. Palomino-Schalscha, *Road Expansion in the Peruvian Amazon*, SpringerBriefs in Latin American Studies, https://doi.org/10.1007/978-3-030-47182-8_8

as unquestionable sources of development, despite their potential to fail (Harvey et al. 2017; Harvey and Knox, 2012, 2015). The political pressure in favour of road construction near isolated protected areas, thus, becomes practically unstoppable in the long term, forcing conservation to step back and make significant concessions (Ráez Luna 2017).

In the case of the Manu province and the Madre de Dios region, the stage seems to have been set for such an effect. The lack of completion of the process of Amazonian 'colonization' of the mid-twentieth century, the absence of the roads that were promised to complete it, left the communities of Andean migrants who joined this project in isolation and made them dependent on the boom and bust cycles of logging (Barbier 2012; Barrantes et al. 2014; Harvey and Knox 2015; Morel 2014). Meanwhile, the establishment of the Native Community framework forced the sedentarization of Amazonian peoples, weakened their territoriality and enclosed their territories to facilitate resource extraction around them (Larsen 2016; Leal et al. 2015; Peluso and Alexiades 2005; Salisbury et al. 2011). In this way, the isolation that Andean migrants faced, and the cultural and economic changes that Indigenous Amazonian peoples went through, gave way to shared aspirations for modernization (Harvey and Knox 2012, 2015; Oliart and Biffi 2010; Ráez Luna 2017; Uribe 2017). However, the possibilities to realize these aspirations were restricted by the establishment of protected areas, which territorialized Manu by prioritizing the global value of conservation over the interests and livelihoods of local communities (Adams and Hutton 2007; Kohler and Brondizio 2016; Oliart and Biffi 2010; Ráez Luna 2017; Shepard et al. 2010). The combination of these factors generated a conflict between the visions of Manu as an area for conservation and Manu as a site of resource exploitation, which remained unresolved for decades, each backed by the political power of their supporters (Gallice et al. 2017; Ráez Luna 2017; Robbins 2004). Peru's adherence to the IIRSA and the resultant arrival of the Interoceanic Highway to the Madre de Dios region in 2011 began to tip the balance in this conflict (Harvey and Knox 2015; Pieck 2013). The region's isolation, which is considered an asset for the conservation of its protected areas, became increasingly eroded by the gradual expansion of the road network (Gallice et al. 2017; Harvey and Knox 2015; Pieck 2013; Shepard et al. 2010). As part of this process, the incremental expansion of the Manu Road, stalled since the 1970s, promises much to local communities but threatens the ACR and the MNP, the 'crown jewel' of Peruvian conservation (Gallice et al. 2017; Larrea-Gallegos et al. 2016; Shepard et al. 2010).

Like the Interoceanic Highway, this unpaved road promises increased speed in travel, political integration and economic connectivity to the communities it is expected to pass through (Harvey and Knox 2012). Consequentially, this road is seen by the people of Manu as the possibility to finally satisfy the overwhelming aspirations of modernity they have held for so long (Harvey and Knox 2012; Kernaghan 2012; Uribe 2017; Wilson 2004). Despite what infrastructures like these promise to bring to these isolated frontier communities, embodied in what we have called the 'road as development' discourse, their apparent benefits are unlikely to be fulfilled given their encounter with 'friction' and the particular conditions of the frontier (Harvey and Knox 2012, 2015; Tsing 2005). However, according to Harvey and Knox, it is exactly this lack of fulfilment which maintains the aspirations of modernity through road connectivity held by frontier dwellers (2012). Our ethnographic research from a Political Ecology perspective shows how the 'enchantments of infrastructure' do take place in the case of the Manu Road. Local testimonies and Eduardo's observations show how off-road communities believe this infrastructure will provide them with benefits that are mostly absent in nearby on-road communities. The only promise this road seems to be able to satisfy is its capacity to increase the profitability of illegal logging, which still has not allowed on-road communities to overcome poverty. These overwhelming aspirations, bred in the isolation and dispossession that frontier communities live in, persist despite their evident lack of fulfilment and are reinforced by several other factors.

Based on detailed ethnographic material and nuanced analysis of the broader context, our research reveals how the presence of protected areas managed by extra-local actors intensifies the capacity of roads to 'enchant'. The accounts and observations that we have presented in this work show how the people of Manu firmly believe that the existence of protected areas in their province has been the very source of their lack of wellbeing (Oliart and Biffi 2010; Ráez Luna 2017; Shepard et al. 2010). Hence, the conservation sector is seen as the great antagonist of local communities in the controversy surrounding the Manu Road, an infrastructure which is seen as the province's escape from stifling isolation. These perceptions seem to constitute a rebound effect produced by the asymmetrical relations between the extra-local agents that manage protected areas and the communities affected by the management strategies they apply (Díaz and Miranda 2014; Larsen 2016; Oliart and Biffi 2010; Ráez Luna 2017; Robbins 2004). This leaves conservation in a politically weak position when faced with the territorializing force of expanding road networks (Kanai 2016; Pieck 2013, 2015). The neoliberal territorial design that IIRSA surreptitiously promotes focuses on serving the global market by providing access to previously inaccessible commodities without ensuring local wellbeing (Kanai 2016; Pieck 2013, 2015; van Dijck 2008; Zhouri 2010; Zibechi 2006). However, the economic growth that the IIRSA network promises extends to its farther reaches and is compatible with the overwhelming aspirations of modernization held in the frontier, supporting the capacity first roads to 'enchant' (Harvey and Knox 2012; Jarrier 2012; Kanai 2016; Zibechi 2006). The combination of these factors increases the capacity of local extractive elites to use the 'road as development' discourse to 'conjure' the benefits that first roads would seemingly provide, which are put in stark opposition to the restrictions that protected areas represent locally (Oliart and Biffi 2010; Ráez Luna 2017; Shepard et al. 2010; Tsing 2005). This increases popular support and political pressure in favour of infrastructures that are most likely to benefit the interests of local elites in greater proportion (Harvey et al. 2017; Harvey and Knox 2015; Tsing 2005). The lack of local support for protected areas leaves the Peruvian conservation sector with few resources in this struggle for land control, which makes resistance to road building ineffective in the long term (Gallice et al. 2017; Ráez Luna 2017). The consequences of this are seen clearly in the continuation of the Manu Road's incremental expansion (Gallice et al. 2017).

Thus, despite the undoubted importance of the conservation of ecosystems, it is still clear that the disenfranchisement and dispossession generated by the strategies through which it is implemented are still counterproductive to its goals in the long term (Kohler and Brondizio 2016; Raycraft 2018; Robbins 2004). Ecosystems as unique as the ones found in Manu province are not only threatened by the degradation and deforestation caused by logging and agriculture, which are its most apparent threats (Gallice et al. 2017; Larrea-Gallegos et al. 2016). They are also threatened in the long term by the territorializing elements of protected areas; the restrictions that these set upon local livelihoods, their failure to provide viable economic alternatives and their lack of recognition of local needs (Gallice et al. 2017; Larsen 2016; Oliart and Biffi 2010; Ráez Luna 2017; Shepard et al. 2010). The fact that protected areas in Peru are an economic liability to the communities under their influence, in monetary and non-monetary terms, is a clear reflection of these conditions (Barrantes and Glave 2014; Díaz and Miranda 2014). Thus, through the research presented here, we have concluded that the continued expansion of the Manu Road is the result of the territorial confinement that communities in Manu face at the hands of the conservation sector and the interests of resource extraction. In this way, the struggle between these territorializing forces, which excludes local voices, constrains the territoriality of the communities of Manu province and reinforces the power of the 'enchantments of infrastructure' near protected areas. Increasing and improving the yet tokenistic local participation mechanisms in the management of these protected areas could bring balance to this situation in the long term (Oliart and Biffi 2010; Ráez Luna 2017; Shepard et al. 2010). However, these mechanisms have little chance of being implemented legitimately until Indigenous knowledges stop being considered as subaltern through the neo-colonial notions that still permeate biodiversity conservation strategies (Adams and Hutton 2007; Bluwstein and Lund 2018; Grosfoguel 2007; Kohler and Brondizio 2016; McEwan 2009; Mignolo 2000; Raycraft 2018). As seen in the testimonies Eduardo gathered, these circumstances pave the way for continued extraction and perpetuate the dispossession and displacement of Indigenous peoples. In a context in which the expansion of the frontier has forced communities into the capitalist system, the short-term profitability of extraction will continue to win over the long-term benefits of ecosystem preservation if the latter provides no viable alternatives for local communities (Murray-Li 2014).

Further exploration of the Political Ecology of the Manu Road and of the other first roads currently being proposed for the Peruvian Amazon will require close attention towards several aspects. The most conspicuous of these will be the consequences of the construction of these infrastructures; the type of development that these actually bring to the frontier, their economic, social and political results, as well as the social implications of their environmental effects. Particular interest should be paid to the way in which local extractive elites, the territories of Native Communities and local social dynamics are affected by this. Furthermore, considering that the Manu Road is now set to finally breach the isolation that the MNP has counted on for so long, a second relevant aspect to consider will be the reaction of the region's conservation agents towards the presence of this infrastructure. The balance between the potential modifications to protective and participative measures will be especially relevant because it can reveal the scope through which Park authorities decide to confront the issue. Moreover, it will be especially interesting to analyse the way in which local aspirations for development change with the presence and continued expansion of the Manu Road and other first roads in the region. A deeper look at the political forces that influence these aspirations and their differences between ethnic groups and within power structures is highly likely to produce further valuable knowledge about the issue of first roads near protected areas in the Peruvian Amazon.

References

Adams WM, Hutton J (2007) People, parks and poverty: political ecology and biodiversity conservation. Conserv Soc 5(2):147–183

Barbier EB (2012) Scarcity, frontiers and development. Geogr J 178(2):110–122

Barrantes R, Fiestas J, Hopkins Á (2014) Evolución de la infraestructura de Transporte y Energía en la Amazonía Peruana. In: Barrantes R, Glave M (eds) Amazonía Peruana y Desarrollo Económico. GRADE, IEP, Lima, pp 109–160

Barrantes R, Glave M (2014) Introducción. In: Barrantes R, Glave M (eds) Amazonía Peruana y Desarrollo Económico. GRADE, IEP, Lima, pp 13–20

Bluwstein J, Lund JF (2018) Territoriality by conservation in the Selous-Niassa Corridor in Tanzania. World Dev 101:453–465

Díaz R, Miranda J (2014) Áreas Naturales Protegidas en el Perú: Efectos sobre la Deforestación y su Relación con el Bienestar de la Población Amazónica. In: Barrantes R, Glave M (eds) Amazonía Peruana y Desarrollo Económico. GRADE, IEP, Lima, pp 209–245

Gallice G, Larrea-Gallegos G, Vásquez-Rowe I (2017) The threat of road expansion in the Peruvian Amazon. Oryx 1–9

Grosfoguel R (2007) The epistemic decolonial turn. Cult Stud 21(2–3):211–223

Harvey P, Jensen CB, Morita A (2017) Introduction: infrastructural complications. Infrastructures and social complexity: a companion. Routledge, London, pp 1–42

Harvey P, Knox H (2012) The enchantments of infrastructure. Mobilities 7(4):521–536

Harvey P, Knox H (2015) Roads: An anthropology of infrastructure and expertise. Expertise, Ithaca, NY

Jarrier A (2012) Seminario Internacional: "Todos los caminos conducen al Brasil. La geopolítica del plan de Integración de Infraestructura Regional Suramericana - IIRSA - en la región andina." In: *Bulletin de l'Institut français d'études andines*, vol 41, pp 303–307

Kanai JM (2016) The pervasiveness of neoliberal territorial design: cross-border infrastructure planning in South America since the introduction of IIRSA. Geoforum 69:160–170

Kernaghan R (2012) Furrows and walls, or the legal topography of a frontier road in Peru. Mobilities 7(4):501–520

Kohler F, Brondizio ES (2016) Considering the needs of indigenous and local populations in conservation programs. Conserv Biol 31(2):245–251

Larrea-Gallegos G, Vazquez-Rowe I, Gallice G (2016) Life cycle assessment of the construction of an unpaved road in an undisturbed tropical rainforest area in the vicinity of Manu National Park, Peru. Int J Life Cycle Assess 11(17/2016):1–16

Larsen PB (2016) Derechos indígenas, gobernanza ambiental y recursos en la Amazonía peruana, Hacia una antropología de la posfrontera. Instituto de Estudios Peruanos, Lima

Leal DB, Salisbury DS, Silva J (2015) Ideas cambiantes sobre territorio, recursos y redes politicas en la Amazonía indigena: un estudio de caso sobre Perú. J Lat Am Geogr 14(2):181–204

McEwan C (2009) Postcolonialism and development. Routledge, London

Mignolo WD (2000) Local histories/global designs: coloniality, subaltern knowledges, and border thinking. Princeton University Press, Princeton, Oxford

Morel J (2014) De una a muchas Amazonías: Los Discursos sobre 'La Selva' (1963–2012). In: Barrantes R, Glave M (eds) Amazonía Peruana y Desarrollo Económico. GRADE, IEP, Lima, pp 21–46

Murray-Li T (2014) Land's end: capitalist relations on an indigenous frontier. Duke University Press, Durham and London

Oliart P, Biffi V (2010) Territorialidad indígena, conservación y desarrollo: Discursos sobre la biodiversidad en la Amazonía peruana. Instituto del Bien Común, Lima

Peluso DM, Alexiades MN (2005) Indigenous urbanization and amazonia's post-traditional environmental economy. Tradit Dwell Settl Rev 16(2):7–16

Pieck SK (2013) Asphalt dreams: road construction and environmental citizenship in Peru. Dev Change 44(5):1039–1063

Pieck SK (2015) "To be led differently": neoliberalism, road construction, and NGO counter-conducts in Peru. Geoforum 64:304–313

Ráez Luna E (2017) In Ráez Luna E (ed), Derechos Ancestrales y Conservación de la Naturaleza en Debate: El caso de los Pueblos Indígenas del Parque Nacional del Manu. Seminario Permanente de Investigación Agraria (SEPIA), Lima

Raycraft J (2018) Circumscribing communities: marine conservation and territorialization in Southeastern Tanzania. Geoforum 100(2019):128–143

Robbins P (2004) Political ecology: a critical introduction. Blackwell Publishing, Oxford

Salisbury DS, López JB, Vela Alvarado JW (2011) Transboundary political ecology in Amazonia: history, culture, and conflicts of the borderland Asháninka. J Cult Geogr 28(1):147–177

Shepard GH, Rummenhoeller K, Ohl-Schacherer J, Yu DW (2010) Trouble in paradise: indigenous populations, anthropological policies, and biodiversity conservation in manu National Park, Peru. J Sustain For 29(2–4):252–301

Tsing AL (2005) Friction: an ethnography of global connection. Princeton University Press, NJ

Uribe S (2017) Frontier road: power, history and the everyday state in the Colombian Amazon. Wiley, New Jersey

van Dijck P (2008) Troublesome construction: the rationale and risks of IIRSA. Revista Europea de Estudios Latinoamericanos y Del Caribe 85:101–120

Wilson F (2004) Towards a political economy of roads: experiences from Peru. Dev Change 35(3):525–546

Zhouri A (2010) "Adverse Forces" in the Brazilian Amazon: developmentalism versus environmentalism and indigenous rights. J Environ Dev 19(3):252–273

Zibechi R (2006) IIRSA: la integración a la medida de los mercados. Ecología Política 31:19–25

The manufacturer's authorised representative in the EU is Springer Nature Customer Service Centre GmbH, Europaplatz 3, 69115 Heidelberg, Germany. If you have any concerns regarding our products, please contact ProductSafety@springernature.com

Printed and bound by CPI Group (UK) Ltd, Croydon, CR0 4YY

30/09/2026

02225474-0005